Laws of Nature, Laws of God?

CW00337545

Laws of Nature, Laws of God?:

Proceedings of the Science and Religion Forum Conference, 2014

Edited by

Neil Spurway

Cambridge Scholars Publishing

Laws of Nature, Laws of God?:
Proceedings of the Science and Religion Forum Conference, 2014

Edited by Neil Spurway

This book first published 2015

Cambridge Scholars Publishing

Lady Stephenson Library, Newcastle upon Tyne, NE6 2PA, UK

British Library Cataloguing in Publication Data
A catalogue record for this book is available from the British Library

Copyright © 2015 by Neil Spurway and contributors

All rights for this book reserved. No part of this book may be reproduced, stored in a retrieval system, or transmitted, in any form or by any means, electronic, mechanical, photocopying, recording or otherwise, without the prior permission of the copyright owner.

ISBN (10): 1-4438-7657-7
ISBN (13): 978-1-4438-7657-5

TABLE OF CONTENTS

Section II: Offered Contributions

Section III: Coda

THE SCIENCE AND RELIGION FORUM

Growing out of informal discussions which began in 1972, around the key figure of Revd Dr Arthur Peacocke, the Science and Religion Forum was formally inaugurated in 1975. Its stated purpose was "to enable and encourage further discussions of the issues which arise in the interaction between scientific understanding and religious thought". These issues, together with the social and ethical decisions demanded by scientific and technological advances, have remained the subject of the Forum's meetings since that date.

In 2005 the Forum merged with the Christ and the Cosmos Initiative. This had been founded by the Revd Bill Gowland, a past President of the Methodist Conference, with the intention of bringing the latest knowledge of scientific thinking within the orbit of the enquiring layperson.

Thus enlarged, the Forum is open to all, of any personal faith or none, who are concerned to relate established scientific knowledge and methodology to religious faith and theological reflection. Implementing its broad objectives, it seeks:

1) to encourage scientists with limited knowledge of religion, and religious people with limited knowledge of science, to recognise and appreciate the contributions of both disciplines to human understanding of life in the world
2) to provide an interface between academics active in science-religion work, and public communicators – notably teachers, clerics, and those training future members of these professions.

At every point, the Forum strives to extend recognition that science and religion, properly understood, are not antagonists, but complementary in the quest for truth.

The Forum holds a regular annual conference, plus occasional smaller *ad hoc* meetings, and publishes a twice-yearly journal, *Reviews in Science and Religion.* Since 2008 it has also published edited proceedings of its annual conferences, under the series title *Conversations in Science and Religion.*

At the date of publication, the Forum's President is Prof John Hedley Brooke (Oxford) and its Chairman Revd Dr Michael Fuller (Edinburgh).

ACKNOWLEDGEMENTS

The Science and Religion Forum warmly acknowledge the highly efficient and outstandingly friendly welcome given to it by the staff of Leeds Trinity University, when they hosted the conference of which these chapters are a product. The architectural and practical qualities of the University's facilities deserve comparable commendation.

As Editor, I owe my own thanks to all thirteen contributors, for meeting a fairly short deadline, and in several instances tolerating a succession of follow-up editorial demands. Particular appreciation is due to Professor Nancy Cartwright, for agreeing to work with an edited transcript of her exhilarating talk as the basis of her chapter, and Dr Jonathan Topham for honouring (splendidly) a commitment which he did not realise, until inconveniently late, that he had implicitly accepted.

I warmly acknowledge also the guidance about various aspects of my Introduction (Part Two), received from Drs Michela Massimi and Fraser Watts, and Miss Danielle Adams.

—Neil Spurway

CONTRIBUTORS

Editor

Neil Spurway studied in Cambridge, but has worked in the University of Glasgow ever since, and is now Emeritus Professor of Exercise Physiology. He has chaired that university's Gifford Lectureships committee, as well as the present Forum, been President of the Royal Philosophical Society of Glasgow and Vice-President of the European Society for the Study of Science and Theology. He co-authored *The Genetics and Molecular Biology of Muscle Adaptation*, but also initiated the present series and edited its first two volumes, *Creation and the Abrahamic Faiths* and *Theology, Evolution and the Mind.*

Invited Contributions

Nancy Cartwright, FBA, currently holds simultaneous Professorships in Philosophy in the Universities of Durham, UK, and San Diego, California. Her immediately previous chair was at the London School of Economics, where she co-headed the project on *God's Order, Man's Order and the Order of Nature*, funded by the Templeton Foundation and run jointly between LSE and San Diego. She is a Past President of both the Philosophy of Science Association and the American Philosophical Association (Pacific Division). Her many books include *How the Laws of Physics Lie, The Dappled World: A Study of the Boundaries of Science* and, most recently, *Evidence: For Policy, and Wheresoever Rigor is a Must*.

John Henry is Professor of the History of Science in the University of Edinburgh. Graduating from Leeds, and obtaining his doctorate from the Open University, he joined Edinburgh's Science Studies Unit in 1986, and has stayed there ever since. His interests include the histories of both science and medicine, and philosophy from mediaeval times to the Enlightenment. Resultant books have been *The Scientific Revolution and the Origins of Modern Science, A Short History of Scientific Thought* and a new translation of Jean Fernel's *On the Hidden Causes of Things (1548).*

Tom McLeish, FInstP, FRS, Professor of Physics and Pro-Vice-Chancellor (Research) at the University of Durham, has previously held academic appointments in Cambridge, Sheffield and Leeds. He has won awards for his research on the molecular theory of complex fluid flow, and currently works on applications of physics to biology, as well as on matters of science policy and history. He gives time to science communication via radio and TV, and the face-to-face contact of schools lectures. His recent book, *Faith and Wisdom in Science* (on which his contribution here is based) has made considerable impact.

Eric Priest, FRSE, FRS, Emeritus Professor in the School of Mathematics and Statistics, University of St Andrews, has devoted his scientific life to theoretical studies of solar physics, and the interactions of magnetic fields with plasmas more generally. He has published two books, more than 30 years apart, on *Solar Magnetohydrodynamics* and *Magnetohydrodynamics of the Sun.* He is committed also to communicating, in lectures, discussions, and both TV and press interviews, his conviction that science and religion are parallel, not conflicting, human quests, with certainty no part of either.

Dr Jonathan Topham is a Senior Lecturer in History of Science at the University of Leeds. His research relates mainly to the history of printed communication in science, especially in late Georgian Britain. Among his co-publications are *Science in the Nineteenth-Century Periodical: Reading the Magazine of Nature* and *Culture and Science in the Nineteenth-Century Media.* He is currently completing a monograph about the Bridgewater Treatises entitled *Reading the Book of Nature: Science, Religion and the Culture of Print in the Age of Reform.*

Shorter Contributions

Paul Beetham, BSc (London), completed a Ph.D in Microbiology in Aberystwyth. Post Doctoral work in Germany, environmental consultation and teaching followed. Entering the Ministry he studied Theology at Durham and is now Superintendent Minister of the Birmingham (West) & Oldbury Methodist Circuit. He has served on the committees of both the Christ and the Cosmos initiative and the Science & Religion Forum.

Geoffrey Cantor is Emeritus Professor of the History and Philosophy of Science at the University of Leeds and Honorary Senior Research Associate at UCL Department of Science and Technology Studies. He has had a life-long interest in figures at the interface between science and religion, and two of his most notable books are *Michael Faraday, Scientist and Sandemanian,* and (with John Hedley Brooke) *Reconstructing Nature: The Engagement of Science and Religion.*

John Emmett recently retired as a Methodist Minister in the Bristol area and Tutor in Christian Doctrine at Wesley College. Before that he worked as a research physicist. He has a BSc in Physics, a PhD in Nuclear Power Physics, both from Imperial College, London, and an MA in Theology and Ministry from Bristol University. He is currently working on a book comparing models used in Trinitarian theology and Quantum Physics.

Richard Gunton, MA (Cambridge), PhD (Leeds), is a Research Fellow in ecology at the University of Leeds. He has he conducted research on plant community ecology in South Africa, Australia, Portugal and France, and contributed to a recent book on *Scaling in Ecology and Biodiversity Conservation.* He is also coordinator of the Faith-in-Scholarship initiative and an advisory board member for the Jubilee Centre.

Gavin Hitchcock (BA, Oxford; PhD, Keele), a mathematician, has worked for a number of years in Africa, mostly at the University of Zimbabwe, where he developed an interest in mathematical talent search and pedagogy. Currently he is Assistant Director (Training) in the South African Centre for Epidemiological Modelling and Analysis at the University of Stellenbosch. His pure mathematical researches have been in topology, and he now writes dialogues in the history of algebra.

John Lockwood, Fellow of the Royal Meteorological Society, a retired Senior Lecturer, University of Leeds, holds a B Sc (Geography with Pure Mathematics) and a Ph D. (Climatology), both from Queen Mary College, London. He researched energy/water exchanges between land surfaces and the atmosphere by numerical modelling. His books include *World Climatology, Causes of Climate*, and *World Climatic Systems.*

Juuso Loikkanen Juuso Loikkanen is a Junior Researcher at the University of Eastern Finland. He holds degrees in mathematics, theology, and economics and is currently working for two PhDs, respectively in Systematic Theology and in Mathematics. His contribution to this book is his first publication in English.

Fabien Revol is Assistant Coordinator, Chair in Science and Religion, and coordinator of the "Jean Bastaire" Chair for a Christian Approach of Integral Ecology in the Université Catholique de Lyon, from which he holds doctorates in both Theology and Philosophy. He is the author of *Le Temps de la Création*, and is writing a book on the question of novelty in nature. He is the only Committee Member of the Science and Religion Forum not based in Britain.

Fraser Watts is Reader Emeritus in Science and Religion in the University of Cambridge, and a past President of the International Society for the Study of Science and Religion. Amongst earlier positions he was Senior Scientist in the MRC Applied Psychology Unit in Cambridge. His publications include books (with co-authors) on *The Psychology of Religious Knowing, Psychology for Christian Ministry* and *Evolution, Religion and Cognitive Science.*

INTRODUCTION

NEIL SPURWAY

Part One: Subject-matter of this Book

As far as we can tell, when science as we now know it took off in the 17th C, every investigator thought of himself as probing some aspect of divine Creation – and every law enunciated was perceived as having been ordained by God for the governance of that Creation. In the more secular ethos of the 21st C, such a position is much less common. In consequence, the philosophical status of "laws of nature" – scientific laws – has become a lot more controversial. The Science and Religion Forum's 2014 conference, in the most congenial surroundings of Leeds Trinity University, was devoted to this topic, and the essays in this book result from talks given at that meeting.

***

The five chapters in **Section One** derive from plenary talks, by invited speakers. In the first, Professor Eric Priest, FRS (St Andrews), intertwines an account of mathematical and experimental studies of the sun with his personal and religious response to the phenomena he encounters – a kind of response which was almost universal among scientists before the so-called "Enlightenment", and is (as he demonstrates) by no means dead today.

In Chapter Two, another FRS physicist, Professor Tom McLeish (Durham) aligns scientific research, over many centuries, with the biblical stance toward nature, especially as expressed in the Book of Job:

> Where were you when I founded the earth? Tell me, if you have insight.

From this he argues that "theology and science" is an inappropriate juxtaposition – we should be striving toward a theology *of* science.

These two splendid essays prepare the ground for what might well be considered the core lecture of the symposium, an account (Chapter Three) by Professor John Henry (Edinburgh) of the theological view of a Law of Nature – a divine injunction to matter as to how it should deport itself – and its subsequent, imperfectly comfortable, secularization. Not only linguistically but, it can be argued, metaphysically the laws discovered by the scientist still carry the implicit connotation of being laws laid down by God, yet the majority of modern practitioners would reject that connotation, and some would question its very meaning.

Chapter Four, by Dr Jon Topham (Leeds) is an account of a series of 19th C lectures, the "Bridgewater Treatises", which were designed to show that the science of the day was wholly compatible with a theistic outlook. It is interesting to consider how the contribution by Professors Priest and McLeish would have fitted, *mutatis mutandis*, into the Bridgewater corpus: I suggest that they would have been amongst the most widely-quoted contributions!

This part of the book ends with a highly-personal philosophical discussion (Chapter Five), by Professor Nancy Cartwright, FBA (Durham and San Diego), of latter-day thinking about scientific laws. This was delivered as the Forum's prestigious Gowland Lecture, open to the public, and is the one contribution to the book which is not an essay re-written entirely after the meeting. We were only able to include a text from this eminent but heavily-committed speaker by transcribing her lecture, and asking her to correct an edited version of that transcript. Accordingly, the printed text includes verbatim elements of the extensive discussion which the lecture aroused. Every other essay embodies, within the continuous written account, such elements of subsequent discussion as the speaker chose to incorporate.

The philosophical level of Professor Cartwright's treatment was sufficiently high that I have judged it appropriate to present an essay of my own, attempting to introduce the modern philosophical thinking about scientific laws upon which she is commenting, for readers unfamiliar with the field. It is embodied in Part 2 of this Introduction.

***

Section Two consists of contributions offered by registrants at the conference. I have placed first (as Chapter Six) a fine essay by Dr Fraser Watts (Cambridge), questioning whether complex biological systems – especially now that they are recognized as embodying massive epigenetic, not just Mendelian influences – will ever be describable by laws, as traditionally understood. Some background to this paper (including a definition of "epigenetic"!) is also included in Part 2, below.

In Chapter Seven Dr Gavin Hitchcock (Stellenbosch) takes us in what might be considered an almost-opposite direction, to consider laws in mathematics. He offers a lovely historical reconstruction to show that, in that discipline, the very concept of law is remarkably recent; and that, insofar as it is now accepted, it has not only an aesthetic but a moral dimension which is hard to detect elsewhere.

The remaining four chapters have more specifically theological themes. As becomes a Frenchman, Dr Fabien Revol (Lyon) builds his paper (Chapter Eight) on the writings of Descartes, and urges the importance of the latter's concept of "continuous creation". This term does not refer to the steady drip-feed creation of new matter, as in the cosmological hypothesis initiated by Hoyle, Bondi and Gold in the 1940s, but to the essentiality of divine immanence, sustaining the Universe in being. It seems appropriate to remark that this was a theme central to the thinking of Rev Dr Arthur Peacocke, the effective founder of the Science and Religion Forum.

Chapter Nine, by Mr Juuso Loikkanen (University of Eastern Finland), argues that there is no need to set divine design and natural causes against each other, and that Intelligent Design, despite raising the hackles of almost every experimental scientist alive, need really not be considered incompatible with their approaches.

In Chapter Ten Dr Richard Gunton (Leeds) raises questions about the validity of "Fine-tuning" arguments for the existence of God. Awe at the lawful appearance of the Universe remains a radically Christian and entirely appropriate stance but, he contends, it would be better expressed without citing "Anthropic" cosmological principles.

Next Dr John Emmett (a retired Methodist minister with a PhD in Physics) contends that dualist thinking, even as suggested by the juxtaposition "Laws of Nature, Laws of God", should be replaced wherever possible by Trinitarian thought-forms. They are, he contends, not only more Christian, but more constructive.

The section concludes, not actually with prayer (as do many religious meetings) but with a chapter *about* prayer. This (Chapter Twelve) is by Dr John Lockwood (Leeds), who considers prayer in the context of weather. As a mathematician, he is well placed to offer some guidance about the mathematics of complex – and even chaotic – systems along the way. Whether such systems will ever be describable in law-like terms it would be rash to predict: they certainly cannot now.

***

Section Three (Coda) consists of just two pieces. The first (Chapter Thirteen), by Professor Geoffrey Cantor (Leeds), was the after-dinner speech at the conference. It returns to the theme of the first two chapters, urging that the only appropriate reaction to the majesty of Creation is reverence.

Finally, Dr Paul Beetham (another Methodist Minister with a science PhD – this time in microbiology) gives an overall, and again very personal reflection on the preceding contributions.

Part Two: Laws in Science

Alister McGrath has said (2005) that what drew him back, from schoolboy atheism towards a religious position, was reading some philosophy of science. This gave him his first awareness of the limits to both the range and reliability of scientific knowledge. Probing just one stage deeper, into the structure of such knowledge as science does provide, one realizes that the nature and role of scientific laws ("laws of nature") has become one of the major questions: and it is, of course, the underlying question of this book. However, we are able to publish here only one chapter dealing frontally with that question from the standpoint of the philosophy of science – Prof Nancy Cartwright's Gowland lecture, asking "How could laws make things happen?" (Chapter Five). This is a brilliant, individualist challenge to a number of widely-held assumptions, but it is not an elementary introduction to the topic! Here, I try to provide such an introduction. It is aimed most directly at sketching some of the background to Nancy Cartwright's chapter, but it should provide an only slightly less direct foundation for parts of several other contributions.

a) Kant and Newton

One talk, given at the conference, cannot be published in this book because it was already in press elsewhere (Massimi, 2014). The speaker, Dr Michela Massimi, is a Senior Lecturer in Philosophy of Science in the University of Edinburgh. Her topic was Newton's conception of natural laws, and Kant's critical reappraisal of Newton's view, undertaken relatively early in his own philosophical life – well before such great works as the *Critique of Pure Reason*. At first glance this early thinking of Kant's might seem to the non-specialist a recherché detail, but it in fact takes one straight to the heart of our conference theme.

In Newton's mind, laws of nature, such as those he himself so majestically expounded, were "laws" in essentially the same sense as laws of the realm – injunctions, prescriptions, given by God to the matter in His creation, dictating how it should behave. That was why they were *called* laws! To this cast of mind, as physical science developed it was making evident the jurisprudence governing the natural world. Newton was by no means the first to think in this way, as John Henry's contribution to this book (Chapter Three) makes clear: Descartes was a particularly important predecessor. But Newton was immensely admired by Kant, whose world-view was substantially founded on Newtonian physics. So for Kant to have detected weaknesses in Newton's outlook upon natural law is striking and significant: essentially, Kant was initiating the modern philosophical debate about the nature of scientific laws, which I presume to sketch below. But before this let me outline the key point which Michela Massimi made about Kant's critique of Newton.

The views she was presenting were those of the pre-1770 Kant, the "pre-critical" Kant, expressed particularly in *The Only Possible Argument* (1763). A key quote is:

> "Something is subsumed under the order of nature if its existence or its alteration is sufficiently grounded in the forces of nature. The first requirement for this is that the force of nature should be the efficient cause of the thing; the second requirement is that the manner in which the force of nature is directed to the production of this effect should itself be sufficiently grounded in a rule of the natural laws of causality."

Reading this attentively, one sees that, for Kant, orderliness, lawfulness, arises "bottom up" from the capacities inherent in nature. It is "sufficiently grounded" there, and not imposed from above, or outside – "top-down" – by Divine will, as Newton's thinking implied. (Historians of Philosophy recognise that Kant was taking the notion of natural "ground" from his older contemporary, Christian Wolff.) Newton's equations brilliantly described what was happening, but his underlying metaphysics involved God too directly for Kant.

b) Modern philosophical thinking about scientific laws

Nowadays, the most basic philosophical suggestion about the nature of scientific laws is that they are, at root, no more than observed regularities. I doubt whether any established philosopher of science actually upholds this "Naïve …" (Armstrong, 1983) or, less pejoratively, "Simple" Regularity Theory (SRT: Bird, 1996), but it makes a good starting-point for such textbook discussions. The SRT is also consonant with the outlook of Francis Bacon, the first writer who attempted to prescribe, at book length, how science should be done. With the overall aim of persuading people to study the world themselves, and not consider it necessary – let alone sufficient – to read what Aristotle had said about a topic, Bacon spelled out the process of generalising from repeating trends in observation, and proceeding on the working assumption that regularities which had been observed so far would continue to apply – an assumption, we may note, which every other living creature, animal, plant or microbe, is continually making too (though this is a 20th/21st C point, not a 17th C one). For this process of generalisation and projection, conducted consciously by human beings, Bacon coined the term "Induction". The inductive search for regularities is clearly an essential first stage in any science, but science is much more than this, and scientific laws are usually more than statements of those regularities.

However, to pin down what more they are is harder. Part of the problem is that there is a diversity of laws, and they do not all do the same things. If we start, as philosophers of science almost always do, with examples from physics, Kepler's Laws of Planetary Motion (the First Law being that the planets move in elliptical orbits, with the sun at one focus of each planet's ellipse) could be regarded as statements of observed regularity: they are wonderfully economical mathematical formulations of the regularities, yet in essence they are regularity-statements, arguably compatible with the SRT. But Kepler's Laws cry out for explanation, for the elucidation of the mechanism which leads to the regularities. That elucidation, of course, was provided two generations later by Newton's Law of Gravitation, building on those of Motion. These are most certainly not just summaries, however elegant, of a massive number of observations. Their statement, verbal and mathematical, implies processes, the intuiting of which constituted huge leaps forward in human understanding. Though Kepler's Laws came first in historical sequence, they are logically derivative from Newton's, and thus secondary to them. Whatever philosophical account we give of one of these groups of laws cannot apply to the other.

Kepler's generalisation from observations and Newton's theorising about mechanism do not represent the gamut of scientific laws. Later, I shall acknowledge some others. However, Kepler and Newton will be enough to have before us as we look at three other philosophical accounts of scientific laws, all more sophisticated than the SRT.

Systematic or Necessary?

The first more sophisticated account was spelled out most fully by David Lewis (1973), following earlier thinking by John Stuart Mill in the 19th C and Frank Ramsey in the mid-20th. Lewis's main concern was to exclude chance regularities from appearing to be laws. He proposed that a regularity represented a law of nature only if it could be construed as an axiom in an overall deductive system which combined simplicity with strength. Simplicity and strength are both, of course, subjective notions, which seems at first sight to be a weakness. However, reading Thomas Kuhn (1962) on the role of training and textbooks in modern science, one must surely recognise that the education of the upcoming scientist these days will align his/her mind to the accepted criteria – it will align subjectivities! So that aspect of Lewis's account, although uncomfortable, is probably true to life. To my mind less comfortable, however, is that the follower of this so-called "systematic approach" must conclude that Kepler's formulations only achieved the status of laws in the wake of Newton's proposals. By contrast, these latter are clearly Lewissian laws in their own right, for they are the axioms of the Newtoniansystem.

Ten years after Lewis's book was published, David Armstrong (1983) used the long-standing philosophical concept of "universals" – properties, such as redness or largeness, which an infinite number of existing or potential things may have in common. For Armstrong, a scientific law is "a necessary relation between universals"[1]. One can get some feel of this in the situation which Bird uses to illustrate it: "If we see various different pieces of magnesium burn in air, we can surmise that the one property, being magnesium, necessitates the other, combustibility." I cannot imagine any practicing scientist gracing this very simple regularity with the title "law", but philosophers, having decided that the detection of regularities must be at the root of all laws, generously apply the term to almost any regularity. Substantively, one must surely wonder what Armstrong's account has added to the regularity that burning is observed in all instances where magnesium is exposed to air? We may intuit necessitation, but its objective meaning is elusive. I am reminded of Hume's critique of the concept of "cause": we may see that B always follows A, but what objective additional information have we added by saying that A *causes* B? Returning to Armstrong: even if we're happy that "necessitation" is meaningful, it is less than obvious how laws such as Newton's law of

---

[1] Briefer but similar analyses had been presented a few years earlier by both Fred Dretske and Michael Tooley.

gravitation can be fitted into the schema: massiveness and distance are universals, but quantitatively particular masses and distances are not, and the counterpoint with particulars is an essential aspect of the traditional concept of what is "universal".

Proponents of necessitation, such as Armstrong, therefore espouse an additional "Principle of Instantiation", whereby universals are instantiated by real particulars in the world. Whether this has done anything more than pinpoint our ignorance, I suspect most of my fellow-scientists will doubt. Other philosophers have doubted too. David Lewis, for instance, insisted (1983) that "necessitation" could not just be postulated: it could not enter into the relations between particulars just by bearing a name, "any more than one can have mighty biceps just by being called 'Armstrong'". (I owe this quote, together with some of the previous understanding, to a widely-helpful overview of which Prof Cartwright was the lead author: Cartwright *et al*, 2005.)

Essentialism

I am content to leave the debate between Systematists and Necessitarians there, because the third account of scientific law is to my mind the most promising. Certainly it is in several ways closer to the way practicing scientists think. Of course, it is expressed in philosophical language, in this case that of Essentialism, a mediaeval concept recalled to serve a modern function. The "essence" of an entity is the sum of its properties or "dispositions" (its "powers", in Nancy Cartwright's terms – Chapter Five). It is through these properties/dispositions/powers that entities interact. Laws such as Kepler's recount our observations *that* they do so, and those such as Newton's our formulations of *how* they do so. Things, out there in the world, have their essential properties, and many of these entail interactions with other things. Modern Essentialism is represented by the work of Mumford (1998), Ellis (2001) and Bird (2008); simpler accounts from the first two authors are Mumford (2005) and Ellis (2008). Essentialist thinking prepares us to accept the conclusion to which Nancy Cartwright comes in her paper here: laws don't make *anything* happen – they aren't causal agents! It's the properties of material entities, and hence the mechanisms to which they contribute, that cause things to happen. What the laws we have formulated do is allow us to predict and explain these happenings.

Explanation

Explanation is, for the scientist, among the crucial roles of scientific law. However, what we mean by "explanation" is also, in philosophers' eyes, not obvious. Half a century ago, the leading account of the role of laws in explanation was that of Karl Hempel, with his "covering law" account (Hempel, 1966, *et prec.*). According to this, a phenomenon is explained if there is a law or laws "covering" the situation, and the observed phenomenon can be deduced from the law(s), together with facts concerning the relevant circumstances – the "antecedent conditions". The occurrence of the phenomenon can therefore be deduced from the law. "Nomological" being philosophical parlance for "having to do with laws", this account is designated the "Deductive-Nomological" (D-N) model of explanation. A variant, encompassing the situation where the law is statistical and the resulting explanation or prediction probabilistic, was an important extension of essentially the same outlook.

For at least two decades after Hempel's view became common currency, discussions of scientific explanation mainly consisted in criticisms of one or another aspect of his D-N model. Among these criticisms was that the model was too tolerant. In cases where the covering law is a symmetrical relation, it does not distinguish between the correct deduction and its inverse. Consider Boyle's Law for a given quantity of gas – that, under conditions of constant temperature, pressure times volume is a constant. The D-N model would give equal justification to a claim that reducing the volume had caused the pressure to rise, and to the claim that increasing the pressure had caused the volume to diminish. If one thrusts inward the plunger of a syringe containing air, the first of these is true; if one transports an air-filled balloon to a greater depth under water, the second applies. They are never simultaneously correct, but the law does not tell us which way round is right. Another instance, perhaps more telling, is the relation between the height of a flag-pole and the length of its shadow: taking Hempel at face value, one could conclude that the length of the shadow caused the pole to be of a certain height! (I owe these examples respectively to Bird, 1998, and Okasha, 2002).

If this aspect of symmetry clearly requires more than Hempel's analysis can provide, another seems satisfying, at least for our purposes here: it is that, in terms of the D-N model, there is symmetry between explanation and prediction: explanation = prediction after the event, prediction = explanation before the event. I think most practising scientists will be happy with that.

More recently, other approaches to explanation have gained credence in the philosophy-of-science community. In particular, the Essentialist view, outlined above, circumvents the problem: according to it, events are brought about by the properties/dispositions/powers of the entities involved. However, while this seems highly persuasive philosophically, and I feel accords well with how my fellow-scientists think at the coal face, an account in terms of laws, and how we use them, is probably still more applicable to formal scientific writing. So the D-N model has not disappeared, and it is referred to again by Nancy Cartwright in Chapter Five, below.

c) Other categories of law in physics

The above impressionistic outline of philosophical thinking about scientific laws was tested against just two examples, the laws of Kepler and those of Newton. They were sufficient to show that no single account can cover the logical status of all scientific laws. A few paragraphs on, I shall begin to ask how well the ideas we have encountered fit the laws formulated in the biological sciences. But we have not quite finished with physics, even at the basic, classical level to which we have so far restricted ourselves.

I first propose a third category of scientific law, in some senses intermediate between the two kinds we have so far acknowledged. The modal instance of this category is Ohm's Law. Ohm's law applies with absolute precision to current flow in electric circuits: precision even greater than that with which Newton's Law of Gravitation applies to the motion of a single planet. It could be documented by an unlimited number of specific instances and, however far one increased the precision of one's instruments, inexactitudes would not be found. But the reason for this is of different kind from those applying either to Kepler's Laws or to the Law of Gravitation. The relation $I=V/R$ can equally well, and more fundamentally, be written $R=V/I$. Algebraically, this is equivalent to the Gas Laws, but the equivalence is misleading. The Gas Laws' terms – pressure, volume and temperature – are defined independently of the laws, which state an empirical relationship between them. But Ohm's law is a *defining* law: voltage and current (rate of flow of charge) can both be independently defined from electrostatics, but resistance cannot. Electrical resistance is *defined* as $R=V/I$. Furthermore, the equivalent relation applies with equal precision to fluid flow in pipes. For "voltage" we now read "pressure", "current" is rate of fluid flow, but hydraulic resistance is, in turn, not merely measured but defined as the ratio of pressure to current.

Yet another category of law is that of Conservation Laws – those of the conservation of energy, momentum, etc.. These are surely different in kind from any of the preceding categories, making it more evident still that no single account can be given of the form or logical status of all laws, even within classical physics? I make no claim to construct a more widely-applicable account of scientific laws, but merely to point out that those so far offered do not look able to embrace all the laws even of their modal science, physics. And science is not coterminous with physics.

d) Laws in the Biological Sciences

> "The philosophy of biology should move to the centre of the philosophy of science – a place it has not been accorded since the time of Mach. Physics was the paradigm of science, and its shadow falls across contemporary philosophy of biology as well in a variety of contexts: reduction, organisation and system, biochemical mechanism, and the models of law and explanation which derive from the Duhem-Popper-Hempel tradition."

This wish (Cohen & Wartofsky, 1976) has not been fulfilled. Physics still dominates the thinking of philosophers about science, as the contributions to this book abundantly demonstrate. But there are a score or more of other sciences. Do laws have similar form in all? We could validly look at representative laws from chemistry, mineralogy or geology, or – more challengingly still – at the sciences of massively complex physical systems, such as geophysics, astrophysics or meteorology – in a degree of counterweight to the customary preoccupation with physics. However, limited for space, I propose to move even further away from physical science, to what one otherwise-admirable review (which shall, at this point, be left anonymous) call "the poor cousins of physics – the special sciences", and consider a sample of biological laws. The ground for this policy is the widely-held suspicion that biological questions differ from those of physics more radically than those of any of the sciences listed a few lines earlier; and if the questions differ in kind, so must the answers.

To start at the least remove, a biochemical or biophysical process may be describable in wholly chemical or physical terms. However, something would be missing from such accounts – namely, the biological significance of the process concerned. As one moves up the scale of biological complexity, through systems physiology, via ethology and psychology to ecology and sociology, this becomes ever more apparent; indeed, beyond physiology, any such attempt would be laughably ridiculous. The argument is closely related to the old, but ever-valid, one which asks whether any understanding of Beethoven's Appassionata sonata is gained by recounting the acoustics of vibrating strings.

So there is something fundamentally different about the subject-matter of biology. Does this mean that the very concept of laws, within biological sciences, is different from that in physical sciences? Discussion of laws, other than those concerning evolution and occasionally also development, is strikingly missing from entry-level textbooks on the philosophy of biology. Exceptions are to be found only in works at the most sophisticated level. To keep things simple, we must start from scratch.

Physiology

Physiology having been my discipline, I take my first example from it, and consider the Law of the Heart, formally enunciated by E.H. Starling, in a prestigious lecture delivered in 1915, though only published three years later. The law states that, over a very wide range, as a heart is more distended during filling so its subsequent contractile force increases, with the clearly desirable result that all the blood which has come in is soon pumped out. If we compare this law with our initial examples from physics, we must surely conclude that, at first sight, it is in the same category as Kepler's laws: they are each summaries of generalisations, crying out for mechanistic explanation. Starling's Law applies to a heart isolated from its nervous control, so it is a consequence of properties residing in the cardiac muscle cells themselves. Work half a century after Starling indicated that it is principally explained by the fact that the contractile force generated by each heart-muscle cell varies with the overlap of two different kinds of protein filament; within the range of lengths over which the cells operate in a healthy heart, the further the filaments are slid apart while relaxed, the greater the contractile force their subsequent interaction produces. Having already noticed the greater parsimony of scientists than philosophers in using the term "law", we should not be surprised that that regularity-statement about interacting filaments is not normally graced by being termed a "law". Nevertheless it surely has the same logical status *vis-a-vis* Starling's law as the Law of Gravitation has towards Kepler's?

However, physiology is only one step removed from biochemistry and biophysics, in the extent to which it seeks to bring physical and chemical thinking to bear on systems of matter which happen to be alive. Perhaps, therefore, there should be no surprise that the status of this representative physiological law seems analogous to that of one from physics.

Genetics

So let us turn next to population genetics, and the Hardy-Weinberg Law, separately enunciated by both its originators in 1908. It asserts that, in sexually-reproducing species, two different forms ("alleles") of a given gene which do not differ in selective advantage will occur in the same proportions in successive generations: "In the absence of selective forces, the gene frequency will remain the same indefinitely". (A human example of difference without evident selective advantage might be blue versus brown eyes.)

The basis of this law is algebraic logic: also, in real terms it is a statistical law, applying with reasonable precision only in large populations. There are two points here. The statistical aspect makes it different from the tiny sample of just three physical laws we have considered so far, but not from Statistical Mechanics, the theoretical analysis which underlies Boyle's (and Charles's) Gas Laws, nor from the empirical laws of Radioactive Decay, and not from countless other laws and law-like generalisations in electronics, astrophysics, meteorology .... Again, the fact that the Hardy-Weinberg formula results from the logical analysis of a conceptual model enables prediction and explanation, *provided* the circumstances fit the conditions assumed in the model, is a common enough situation in physics, too. The Hardy-Weinberg Law sets up a null hypothesis – no evolution – which, by its almost universal failure to tally with observation, demonstrates that there is hardly any real population which is not constantly evolving (Keeting, 1980); and the detailed respects in which a particular set of observations departs from the algebraic prediction give a good indication of the nature of the evolutionary trend in question. In terms of the SRT, the simple regularity this law describes is that of a theoretical model with which observations almost always fail to fit: it is virtually the opposite of Baconian induction! But this does not make it different in kind from model-based laws in the physical sciences. In classical physics, the Ideal Gas Laws, and Newton's Law of Cooling, have comparable relations to the real world; and in current research fields, from particle physics to cosmology, such situations are commonplace.

However, Starling's and the Hardy-Weinberg law still have one major feature in common: that the mechanisms underlying each can be analysed in material, essentially-molecular terms. In the first instance, these are the interactions of cytoplasmic proteins, in the second the trajectories of genes through the processes of cell division. As a last biological example, let us look at phenomena which are not objective but subjective, and the fundamental data are the reports of conscious human beings – psycho-physics – and the search for regularities in the intensity of sensation.

Psycho-physics

The original law, enunciated mathematically by G.T. Fechner in 1862, states that the experienced difference between the magnitudes of two stimuli varies as the ratio *(not* the arithmetical difference!) of the stimulus intensities. It follows that **S**, the experienced intensity of a sensation = **k.logR**, where **R** is the physical intensity of the stimulus and **k** a constant for the particular sense modality concerned. Thus to match the sensed effect of turning on a second light bulb (of the same wattage) when initially there was one, if we start with 50 bulbs we must turn on, not one more (the same arithmetical increment), but another 50. The equivalent has been found to apply, more or less closely, to at least thirty different sensory modalities, including sounds, smells and tactile pressures. And the logarithmic ratio is enshrined in the decibel scale by which engineers indicate sound intensities.

The story of this law, since its original statement, is not one of steady maturation but of challenge and revision. Fechner based his mathematics, as Weber before him had based the pioneering experiments, on minimum detectable increments – "just noticeable differences". When experimenters such as S.S. Stevens, in the mid-20th C, focused on subjects' judgements of "half", "double", etc., the strength of a primary stimulus, they found that the best approximation to a law was more accurately stated mathematically not as a logarithmic but as a power relation, $\mathbf{S} = \mathbf{kR}^{\mathbf{x}}$ where **x** varied rather widely with different modalities. In the medium range of intensities the values indicated by this kind of equation are often not very different from those predicted by the logarithmic law, but where the stimulus intensities differ widely the two formulations are far apart.

If we return to our comparisons with laws in physics, it seems clear that even the move to subjectivity has only reduced the precision of the formulation, it has not altered anything in principle as regards the possibility of discerning regularities and expressing them in laws. As regards explanatory potential, Stevens' law is no more mechanistic than Fechner's; each seeks to encapsulate the regularity, not to explain it. Nevertheless the mechanisms are coming to be understood, and evidently reside in molecular and electrochemical phenomena in the membranes of the various sensory and neural cells. The situation is closely comparable to the cardiac one, where Starling's law as such merely encapsulates the regularity, though subsequently we have seen that it can be explained molecularly. The Fechner, Stevens and Starling laws are all, therefore, at what we might call the Kepler level, not that of Newton, let alone Einstein. Yet even in this respect the honour of biology can be saved, for the Hardy-Weinberg law was derived precisely by theoretical consideration of the behaviour of the genetic particles postulated by Mendel. The word "gene" had yet to be coined, and the double helix was still two generations in the future, but the mathematics of Hardy and Weinberg embodies a mechanism just as much as Newton's did.

The Evolutionary Perspective

I have argued, therefore, that there is no fundamental difference between the physical and the biological sciences, as so far examined, in respect either of the feasibility of encapsulating observed regularities in formal laws or of those laws being capable of being derived from an understanding of mechanism. Yet to my mind, and I believe those of the overwhelming majority of current biologists, there remains a respect in which all biological laws – those considered above, and the many others we have not sampled – differ fundamentally from all in the physical sciences, whether physics itself or any other in the range from chemistry to cosmology. Virtually every mechanism described by a biological law applies because it has enhanced the capacity, of the animals, plants or bacteria which display it, to survive and thence to reproduce. The Hardy-Weinberg law refers directly to reproduction, the other laws to the individual survival necessary for reproduction. Thus Starling's Law demonstrates an elegant functionality in ensuring that blood circulates without wasteful accumulation in either the venous or the arterial system. He demonstrated it directly in dogs, but it, or something close to it, almost certainly applies to all animals which have hearts. And the laws of sensation, whether in Fechner's or Stevens' formulation, represent wonderful economy: the fine discrimination necessary for sensitivity to modifications in gentle stimuli would be astronomically wasteful at the high-intensity end of the response range: wasteful, that is to say, in terms both of the numbers of sensory receptors and nerve fibres required and the metabolic energy they would cost. Exactly equivalent laws apply to the nerves controlling muscles: finesse of control is highly desirable at the low-force end of the muscle's range, but would be hopelessly wasteful at the upper end.

I conclude that the concept of adaptive evolution is truly fundamental to all biological thought, and underlies all biological laws. In the much-quoted dictum of Theodosius Dobzhansky, "Nothing in biology makes sense, except in the light of evolution". It is not in their individual form, but in their unanimous embodiment of that underlying vision, that the laws of biology, of the kinds we have considered so far, differ fundamentally from those of physical science.

e) Evolution, Epigenesis and Complexity

I remarked near the beginning of Section (**d**) that entry-level books on the philosophy of biology make no reference to the sorts of law considered in that section, but all consider a range of problems which philosophers have claimed to find in evolutionary theory itself. Given the importance which I have just allocated to the evolutionary concept, I can clearly not question the appropriateness of the philosophers' interest, though I confess to considering most of their concerns misplaced! However, no contributor to this book addresses these issues, so I must not air my criticisms here.

One contributor, however (Dr Fraser Watts, Chapter Six), considers another major problem which modern biology raises for the philosophy of science – that of seriously complex systems (where "complex" means more than merely "complicated"). The kinds of system at issue here are not individual hearts, but at very least the whole cardiovascular system, and more often the entire animal under stress; not the distribution of alleles in successive generations of a single species but the development over time of the competing populations in a complete ecosystem; and not the response of one sensory modality to variations of stimulus intensity, but that of the whole brain to its endlessly changing environment. These are massive challenges to biological theorists, and it is hard to imagine their work ever being satisfactorily treated by the kind of philosophy of science outlined earlier. The possibility of embodying such thinking in a deductive-nomological model, for instance, seems remote. Perhaps Nancy Cartwright's "powers", or the related Essentialism of other thinkers, can accommodate such problems linguistically, but I am as doubtful as is Fraser Watts that doing this will ever add significantly to our scientific understanding.

Dr Watts is not, of course, the first to have recognised the philosophical, as well as scientific, challenge of complex biological systems. Ayala & Dobzhansky (1974), Maturana & Varella (1980), Bechtel & Richardson (1993), and Mitchell (2003) are representative of those who wrote in earlier decades on this topic. One respect in which Fraser Watts' chapter is invaluable alongside these predecessors is that he makes his case with remarkable simplicity and clarity – not always the case in the prior literature! Another is his specific, and very telling, theological orientation.

But the third respect is that he takes full account of a marked change in the tenor of recent thinking about the actions of genes, which even the most recent of the above authors (Mitchell) seems, as I read her, to have completely failed to spot. The term to conjure with is "epigenetics". I will not duplicate the wonderful selection of references which Fraser himself cites, but it may be helpful if I allow myself a paragraph presenting this concept, in my own terms, to readers who have not met it.

Let us start with the genetic law, from 1908, of which I have made extensive use above, the Hardy-Weinberg law. This is a key edifice of simple Mendelian genetics, in which one gene produces one effect in the developed organism – "bean bag genetics", as Ernst Mayr has disparagingly described it. This genetics reached its apogee in the work of Watson and Crick, published in 1953. Their "triplet code" of base pairs in DNA, specifying the production of a single protein by the cells' metabolic processes, dominated biological thought for the ensuing generation. By the 1980s, however, it was encountering anomalies at an unignorable rate. In terms of Thomas Kuhn's take on the history of science (1962), the situation was ripe for a further revolution (Hewlett, 2011). A large percentage of DNA (up to 95%, we would now say) did not seem to work in the Mendel/Crick-&-Watson way, and was dismissed for a while as "junk DNA". But it isn't junk, it does something different and subtler than was then appreciated: it modifies the action of the simple Mendelian genes, usually in response to differences in the cell's or the whole organism's environment. Developmental stages, seasonal changes, influences from the surroundings, and many other agents can all affect the resultant organism via these modifying genes. The science involved is no longer just genetics, but (in the 1942 coinage of C.H. Waddington) "epigenetics". Developmental biology, a difficult enough subject anyway, has become ecological and evolutionary developmental biology, "eco-evo-devo" (Gilbert & Epel, 2009). This change is so radical that nothing written without recognizing it can now be considered to give anything better than a very partial account of biological complexity.

Finally however, as Fraser Watts acknowledges, there are complex systems which are confronted by the physical sciences as well as the biological, though, currently at least, they are much fewer and, obviously, they embody no aspect equivalent to epigenesis. The key instance, to which Fraser himself alludes, is weather, and this theme is enlarged upon by one other of our contributors, Dr John Lockwood (Chapter Twelve). It is good that these partially parallel challenges in biological and physical science are both represented in this symposium. Perhaps there is scope here for a whole conference of the Science and Religion Forum?

Meanwhile, it is well to be reminded that, if ever our thinking seems for a while to be coming abreast of the world's multiplicity, it will pretty soon prove mistaken.

f) Kant, Newton and their latter-day successors

It is fitting finally to return to the question over which Kant took issue with Newton – the theistic perspective on natural laws. In addition to being the theme of John Henry's essay, this question is touched upon, in various ways, by most other contributors to this volume. It was also the subject of the winning entry in the Peacocke Prize, the Science and Religion Forum's annual essay competition for students and young professionals, run in association with the conference. This year's winner, Miss Danielle Adams (Leeds), was present at the meeting, and referred to the topic (though not to her essay: Adams, 2015) in discussion ("Questioner 11"). In these concluding paragraphs I attempt to bring Newton, Kant, Lewis, Armstrong, Cartwright, the Essentialists and Adams all together.

The view, termed by Adams the "Sovereignty thesis", which Newton shared with Descartes and other predecessors, namely that God not only created the natural world but ordained the laws by which it must behave, implies that logically, if not temporally, matter precedes those laws. To the philosopher, the laws as Newton expressed them were thus "contingent" – they could have been otherwise. This is clearly compatible with Armstrong's "top-down" metaphysics: however elusive "necessitation" may be from the scientific standpoint, the theist cannot fail to see it as originating in God, and imposed by God on matter. The position of sophisticated regularity theorists, such as Lewis, must also be adjudged a "top-down" one, for laws, on their account, embody not just the observed regularities but the "system" in which the regularities are embraced. Both positions imply that it could logically have been the case that the entities we see in the world existed yet obeyed different laws; the theistic take on this must be that the laws we actually observe are those willed by God.

By contrast, Kant's "bottom-up" view of 1763 leads easily to an Essentialist account, in which the behaviour of natural entities reflects their properties, and natural laws simply generalize the consequences. Nancy Cartwright speaks below of "powers" and "mechanisms", rather than "properties" or "dispositions", and certainly doesn't refer to "essences", but the broad pattern of her thought appears to be much closer to Kant than to Newton. So where Adams concludes that God's creative process is of material entities which have inherent properties, and the laws of science encapsulate the implications that human investigation has so far identified, I like to think she is writing in a lineage which includes both Kant and Cartwright, as well as Ellis, Mumford and Bird.

References

Adams, D. (2015) "God and Dispositional Essentialism: An Account of the Laws of Nature" (*submitted for publication*). For personal perusal, access:
http://www.leeds.ac.uk/arts/profile/20042/1059/danielle_adams.

Ayala, F.J. & Dobzhansky, T. (1974) *Studies in the Philosophy of Biology*. London & Basingstoke: Macmillan.

Armstrong, D. (1983) *What is a Law of Nature?* Cambridge: Cambridge University Press.

Bird, A. (1998) *Philosophy of Science*. London: UCL Press.

—. (2008) *Nature's Metaphysics: Laws and Properties*. Oxford: Oxford University Press.

Carroll, S.B. (2006) *Endless Forms most Beautiful*. London: Weidenfield & Nicolson (p/b edn London: Phoenix, 2007)

Cartwright, N. & Alexandrova, A., with Efstatiou, S., Hamilton, A. & Muntean, I. (2005) "Laws" in F. Jackson & M. Smith (eds), *The Oxford Handbook of Contemporary Philosophy*. Oxford: Oxford University Press.

Cohen, R.S. & Wartofsky, M.W. (1974) *Methodological and Historical Essays in the Natural and Social Sciences*. Dordrecht: Reidel.

Ellis, B.D. (2001) *Scientific Essentialism*. Cambridge: Cambridge University Press.

—. (2008) "Essentialism and natural kinds" in Psillos, S. & Curd, M. (eds), *The Routledge Companion to Philosophy of Science*. London & New York: Routledge.

Gilbert, S. & Epel, D. (2010) *Ecological Developmental Biology*. Sunderland, MA: Sinauer.

Hempel, C. (1966) *Philosophy of Natural Science*. Eaglewood Cliffs, NJ: Prentice-Hall.

Hewlett, M. (2011) "'Is it Tomorrow, or just the End of Time?': Paradigm Shifts in the Biological Sciences." In Navarro, J. (ed.) *Science and Faith within Reason*. Farnham, UK: Ashgate.

Kant, I. (1763) "The Only Possible Argument in Support of a Demonstration of the Existence of God" (in German). *Akademie Ausgabe, BDG* vol 2.

Keeting, W.T. (1980) *Biological Science*. New York & London: Norton.
Kuhn, T. (1962) *The Structure of Scientific Revolutions*. Chicago: University of Chicago Press (2nd edn 1970).
Lewis, D. (1973) *Counterfactuals*. Oxford: Blackwell.
—. (1983) "New Work for the Theory of Universals". *Australian Journal of Philosophy*, 61, 343-77
Massimi, M. (2014) "Prescribing laws to nature", *Kant-Studien,* 105, 491-508.
Maturana, H., & Varela. F. (1980) *Autopoiesis and Cognition: The Realization of the Living*. Boston: Reidel
McGrath, A. (2005) *Dawkins' God: Genes, Memes and the Meaning of Life.* Oxford: Blackwell.
Mitchell, S.D. (2003) *Biological Complexity and Integrative Pluralism.* Cambridge: Cambridge University Press.
Mumford, S.D. (1998) *Dispositions*. Oxford: Clarendon Press.
—. (2005) "Kinds, essences and powers", *Ratio*, 18, 420-36.
Okasha, S. (2002) *Philosophy of Science: a Very Short Introduction.* Oxford: Oxford University Press

SECTION I:

INVITED CONTRIBUTIONS

CHAPTER ONE

A SCIENTIST'S VIEW OF THE LAWS OF NATURE AND OF GOD

ERIC PRIEST

Introduction

I am a mathematician who applies his mathematics to try and understand the Sun, but I am also a Christian. Indeed, both parts of my life have involved journeys of discovery. My plan in this article is to describe briefly what studying the Sun has entailed and then to make some general comments about the nature of laws of science. Then I shall give broader comments about what we know about the big bang, before concluding with some personal comments on the interaction between science and faith.

Our Sun

Some solar physicists study the Sun because of its profound influence on the Earth, while others do so because of its central importance for astronomy as a whole, since it is a laboratory where we can investigate fundamental laws of the cosmos in much greater detail than anywhere else in the Universe.

However, my motivation is simply that I am fascinated by the amazing structures and dynamic processes that occur around, on and inside the Sun. Modern science reveals a quite extraordinary Universe, much stranger than fiction, and often gives us a sense of mystery and beauty.

The Sun is a ball of gas held together by gravity. Its interior is opaque, while its atmosphere is transparent, and so radiation can escape from the surface of the Sun (Fig. 1-1) as most of the light that we see with our eyes. In the surface you can also observe dark regions called *sunspots*. Sunspots represent regions where very strong magnetic fields are coming through the surface and spreading out to fill the whole solar atmosphere.

All of the structure and dynamic behaviour in the atmosphere is created by the magnetic field. However, the Sun is not a normal gas. We are all familiar with the three states of matter that surround our everyday life – solids, liquids and gases – and they can go from one state to another by increasing the temperature. Heat a solid (such as ice) and it becomes a liquid;

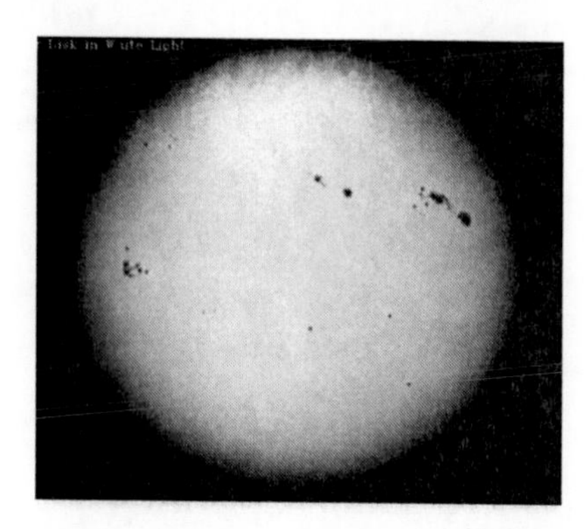

Fig. 1-1. The Sun's surface in normal white light including a few sunspots (courtesy ESA/NASA SoHO Mission Teams)

then heat water and it becomes a gas (water vapour); but if you continue raising the temperature of any gas it eventually turns into the fourth state of matter, which we call *plasma*. A plasma is an ionized gas, where the atoms of a normal gas have been split up into positive ions and negative electrons.

The whole of the Sun is plasma, which behaves quite differently from a normal gas in one respect, namely, that it is coupled in an intimate and subtle way to the magnetic field, and it is this coupling that causes much of the dynamic behaviour that can be observed.

If you go up in the Earth's atmosphere, you eventually reach the ionosphere, and that is where the atmosphere has turned into the plasma state. Indeed, the whole of space between the Earth and the Sun is plasma, as is the whole of the solar system and the Galaxy, reaching out to the edge of the visible Universe, with the exception of a few lumps of solid (planets, comets and so on).

The Corona

The outer atmosphere is called the *corona* and can be viewed during a total solar eclipse (Fig. 1- 2). The solar surface is only 6000 degrees Kelvin (K),

Fig. 1-2. The solar corona during a solar eclipse (Courtesy High Altitude Observatory, Boulder, Colo.,USA)

but it was discovered in 1940 that the corona has a staggeringly high temperature of several million degrees. It is well established that it is the magnetic field that is heating the corona, but the precise mechanism has not yet been identified.

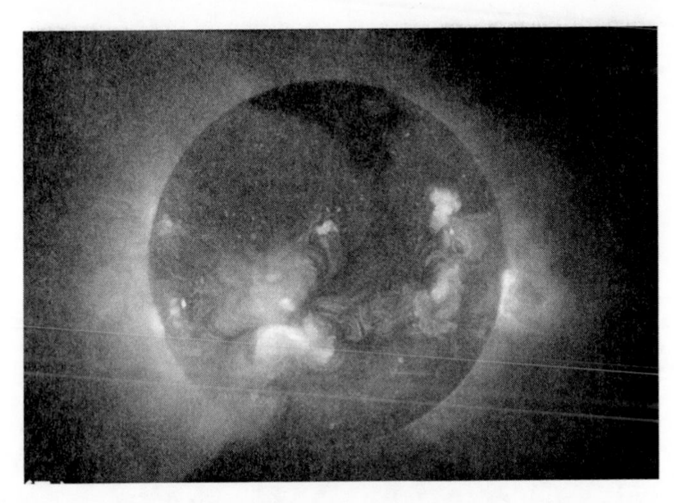

Fig. 1-3. The corona viewed in soft x-rays by the Hinode satellite (courtesy Hinode/XRT team)

The corona can also be observed direct without waiting for an eclipse by imaging the x-rays coming from the Sun (Fig. 1-3). Objects glow with different colours depending on their temperature. As the temperature increases, the colour becomes bluer, and when the temperature is a million degrees it has become so far beyond blue that it glows in x-rays. No x-rays come from the solar surface, and so that is why there is no need to wait for an eclipse. All of the beautiful plasma structures seen in Fig. 1-3 are created by the magnetic field.

The power of modern observations from space satellites is amazing. For example, the left of Fig. 1-4 shows the whole disc of the Sun in normal white light from the Japanese-UK-USA satellite called Hinode.

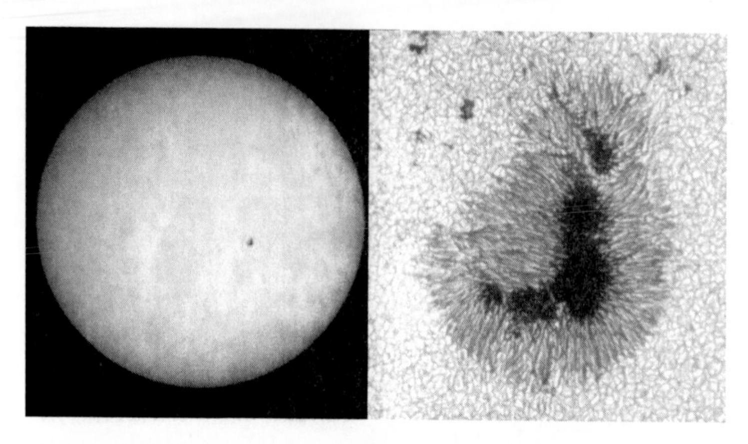

Fig. 1-4. The whole Sun (left) and (right) a close up of a sunspot (courtesy Hinode/SOT team

Just to the right of centre you can see a sunspot, and on the right is a close-up of that sunspot, revealing an incredible amount of dynamic fine-scale structure that we are in the process of trying to understand.

$$\frac{\partial \mathbf{B}}{\partial t} = \nabla \times (\mathbf{v} \times \mathbf{B}) + \eta \nabla^2 \mathbf{B} \quad \text{Induction Equation}$$

$$\frac{D\rho}{Dt} + \rho \nabla . \mathbf{v} = 0 \quad \text{Mass Conservation}$$

$$\rho \frac{D\mathbf{v}}{Dt} = -\nabla p + \mathbf{j} \times \mathbf{B} + \rho \mathbf{g} + \textbf{Viscous Terms} \quad \text{Motion}$$

$$\frac{\rho^\gamma}{\gamma - 1} \frac{D}{Dt}\left(\frac{p}{\rho^\gamma}\right) = \nabla . \left(\kappa_\parallel \nabla T\right) - \rho^2 Q(T) + H(s, t, \mathbf{B}, \rho, T) \quad \text{Energy}$$

$$p = \frac{R \rho T}{\mu} \quad \text{Gas Law}$$

$$\nabla . \mathbf{B} = 0 \quad \text{Gauss' Law}$$

Fig. 1-5. Equations describing magnetic field-plasma interactions

The mathematical equations that describe the interaction between the magnetic field and plasma in the solar atmosphere are, for fun, shown in Fig. 1-5. They are called magnetohydrodynamic equations and represent a unification between the equations of electromagnetism and those of fluid mechanics. To me these equations are incredibly beautiful – they have intrigued me for many years, but I am sure they possess many more subtle secrets that are just waiting to be discovered – truths about fundamental plasma behaviour in the universe.

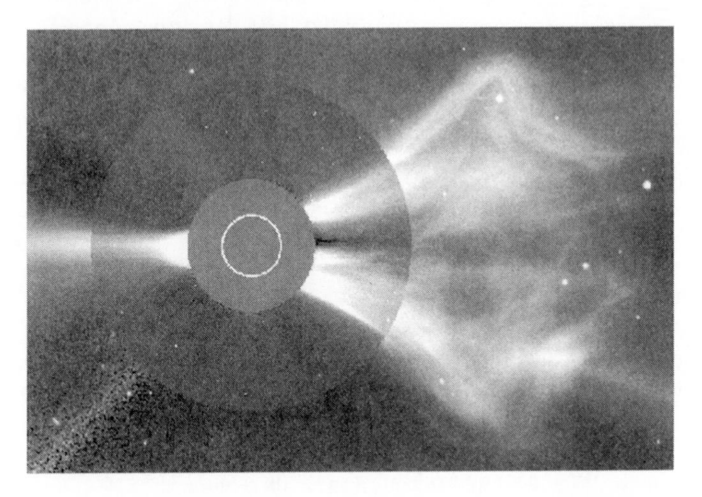

Fig. 1-6. A coronal mass ejection (courtesy SoHO/LASCO team)

When a highly complex group of sunspots appears on the Sun, there is an air of excitement in the solar community as we wait for an enormous solar flare to take place, accompanied by a huge ejection of plasma and magnetic field called a *coronal mass ejection* (Fig. 1-6). When such ejections reach the Earth they can produce beautiful displays of northern lights (the aurora borealis), but they also affect the Earth's space environment in many other ways, such as damaging space satellites, creating cell phone dropouts and producing high-energy radiation which encourages airplanes to avoid polar routes.

Another impressive natural event is a *solar eclipse*, which occurs each year somewhere on the Earth and once in 400 years at any one location. A total eclipse is present when the whole of the Sun's disc is covered by the Moon as viewed from the Earth; this takes place along a narrow track across the Earth's surface. Over one and a half hours, as the Moon gradually covers up the Sun, the light level at the Earth's surface slowly decreases, but the Sun is so bright that you have to look at it through specially darkened glasses; the air becomes cooler and the shadows sharpen. In the last ten minutes the air of excitement increases, as the light level and temperature decrease more quickly, and you feel a wind as air moves in towards the cooling region. Also, the planets appear in view, stretched out in a line, either side of the Sun, since they all orbit the Sun in a plane.

Then, all of a sudden, you see the corona glowing white and possessing beautiful shapes stretching far away from the solar surface (Fig. 1-2). The corona has the same brightness as a full Moon and so you now can look at it with your unprotected eyes. Looking around the horizon, there is a sunset in all directions. Totality lasts for a few minutes as you gaze in wonder at the corona, the most impressive of all natural events on Earth. Then, as the Moon moves on, you start to see sunlight shining though craters in the Moon's surface, followed by the so-called diamond ring effect. As the brightness rapidly increases again, you have to turn away, and you experience a strong desire to turn the clock back and view the wondrous event again.

For me, studying the Sun gives a sense of wonder and of beauty, partly because of the structures we observe and partly because of the underlying laws that we discover. Indeed, in my research I am trying to uncover laws of basic plasma behaviour: How is the magnetic field generated inside the Sun? What is the internal structure of the Sun? How is the solar wind accelerated? Why do huge plasma eruptions occur? How do solar flares convert magnetic energy in the corona into other forms?

The growth of modern science was stimulated by the Christian doctrine of creation, which suggested that the Universe is orderly, good and worthy of study. Why is it that mathematical laws appear to underpin the Universe? One possibility is that they are evidence for a divine Creator, that they flow from His being. Another is that they are independent of God and that there is no need for a creator.

To me the first possibility is more satisfying for three reasons:

(i) It is uncanny that the mathematical symmetry and elegance mirrors the patterns we see in nature, suggesting a deeply rational heart to the Universe. The mathematics is no human invention but represents a deep reality behind the world, reflecting the divine mind with its uniformity, regularity and intelligibility.

(ii) The following deep questions need answers: Why are there mathematical laws at all? Why is nature mathematical in form? Why are the laws universal? Why is there something rather than nothing?

(iii) The mathematical laws give no explanation for themselves. No complete mathematical system contains a proof of its own consistency, since there are always propositions that are unprovable (Goedel's theorem).

Laws of Nature and of God

Is a Given Law True? Does God exist?

In pure mathematics you can prove that results are true, but these aspects of truth are always very limited and specialized. In applied mathematics and in most of science, on the other hand, you cannot prove a theory is correct. But what you can do is ask whether it is consistent with your observations.

In a similar way, Christianity can never prove that God exists, but it can ask whether His existence or non-existence is more consistent with our experience. For me, God's existence is much more consistent, and so I am happy (for the time being) to live my life under that assumption – to try and follow the example of Jesus, to study the bible, to be part of the Christian community and to listen to the promptings of the Holy Spirit.

In most of science, the laws of nature that you discover are often provisional, valid in certain regimes, approximations to, or windows on, absolute reality. Occasionally, there are major paradigm shifts, where the scientific community realizes the limitations or invalidity of the previous laws. For example, Newton's second law of motion was revolutionary at the time (1687) and remained consistent with observations for 200 years. It is descriptive rather than prescriptive and so is not an absolute law. Today it is still consistent with observations, provided speeds are much smaller than the speed of light and length-scales are much larger than the so-called de Broglie wavelength otherwise it is replaced by either general relativity or quantum mechanics.

However, there has been a paradigm shift in the underlying philosophy. For Newton, the Universe was governed by universal and fixed laws of nature that reflected a deep truth about the nature of reality and led to the popularity of deism in the 17th and 18th centuries. For Newton, space and time are absolute and the apple falls because of God's will – he makes its existence and movement possible. Furthermore, although within Newton's equations time can run in either direction, a possible explanation for the forward arrow of time was later given by Ludwig Boltzmann using his second law of thermodynamics.

For Einstein's relativity theory, space and time are no longer absolute and gravity is not a force but is due to the curvature of space-time.

To Augustine, the idea of a time before creation was meaningless since God is outside time. But, if God is outside time and experiences a block view, seeing all of space and time at once, how does God interact with time? What about free will? Maybe, also, there was indeed a time before the big bang. Thus, the whole question of God's relation to time is a puzzle – does He exist in other dimensions? Is time contingent upon God?

Quantum mechanics is even stranger than relativity theory. According to the standard Copenhagen interpretation, an observer affects an observation and there is no independent reality. The irony here is that science started by trying to explain the nature of reality and has ended up so far with quantum mechanics, where reality is illusive.

What is the Relation between God and the Universe?

Four possible relationships between science and religion were discussed by Ian Barbour (1997), namely that they are in conflict, independent, in dialogue or integrated.

I have personally never felt a conflict between science and religion; conflict arises only if you misunderstand the nature of one or the other. Thus, a fundamentalist view of Christianity, a wooden literalism, is at odds with history – for example, St Augustine (400 AD) said "you should not interpret scripture in a way that conflicts with reason or experience". On the other hand, scientism's claim that "science answers every question" is also clearly false.

So what is the relationship between God and the Universe? In the ancient pagan world, there were many gods who occasionally interfered in the world, and this led ancient philosophers to propose two main accounts of reality. The Stoics suggested that divinity pervades the whole cosmos, with the *logos* being the rational principle of order built into the universe, so that God is essentially everywhere.

On the other hand, the Epicureans thought that, if the gods existed, they took no interest in the world. This led on to Stephen Jay Gould's (2002) idea of *non-overlapping magisteria*, with a clear separation between science and religion (Fig. 1-7). There is, in this view, no interaction between them, so that they can each get on with their own realms of study without conflict. This in turn tends to lead on to the idea of a deist God, remote and uncaring, or eventually to the extreme of doing away with religion altogether as science increases its range (scientism).

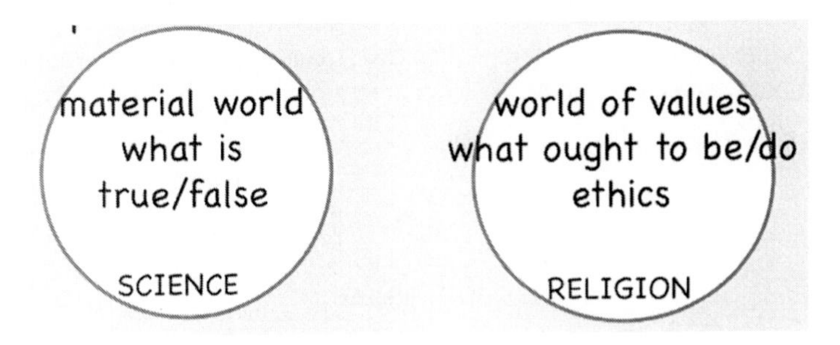

Fig. 1-7. One view of the relation between science and religion

However, a third view is more appealing to me (Fig. 1-8), namely, of a fuzzy boundary or interaction region, where you can ask different questions about the same reality. However, these are immersed in an underlying unity of what it means to be human, including ideas of creativity, community, beauty and wonder.

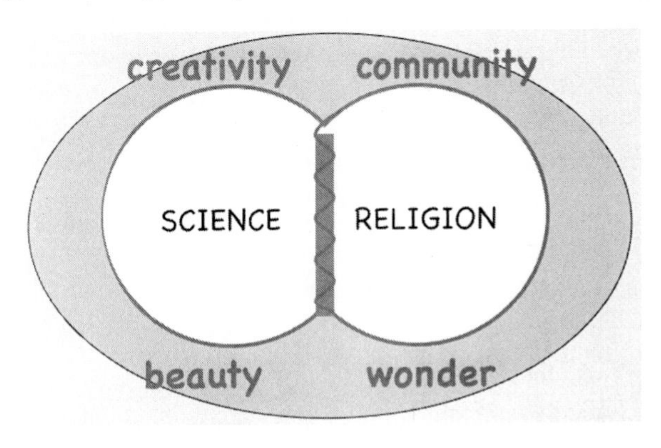

Fig. 1-8. An alternative view of the science-religion relation

An integrated approach, however, appeals best to me, and has been illustrated in a parable that Hyung Choi once showed me. He considers the two traditionally separate islands of science and religion (Fig. 1-9a), each with its own terminology and language, which some people attempted to link by a bridge. However, a closer look at the islands shows them to be located in a mist, and when the sun comes out the mist clears to reveal that there is no sea at all between the islands, but solid ground that had been obscured by the mist (Fig. 1-9b). So perhaps there is one island, one underlying reality – one building but different drawings or maps – one truth illuminated from different directions. But each map or direction is incomplete, since all are needed for a fuller understanding.

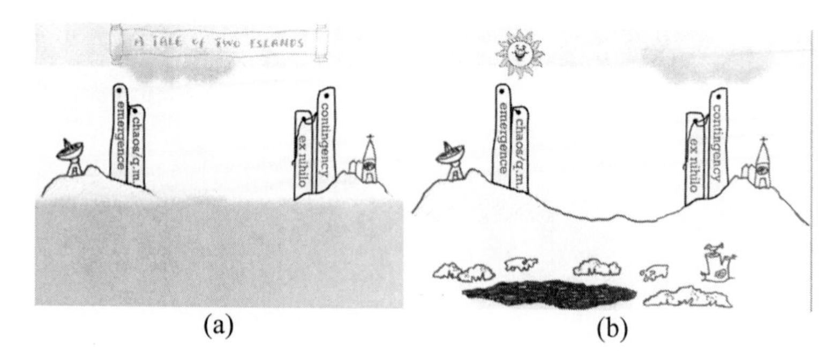

Fig. 1-9. A tale of two cities by Hyung Choi (2004)

This fourth idea is close to the recent proposal of Tom McLeish (2014), who suggests that all topics should be open to discussion by science or religion. For him, we should work in future to develop a theology of science as well as a science of theology.

What is the Nature of Science-and-Faith Questions?

For any question we should ask 'is it scientific or non-scientific?' Both are clearly important, but whereas 'is the Earth warming?' or 'how did *homo sapiens* arise?' are scientific questions, 'what should we do in response to climate change?' or 'does God exist?' are non-scientific.

We can ask different questions about the same event. Thus, 'how is a kettle boiling?' will involve chemistry and physics, but 'why is it boiling?' may be a response to my wife's thirst. In other words, mechanism does not exhaust meaning. Another example would be 'why or how are two people kissing?' Why may be answered in terms of their relationship, but how may produce the dry answer 'the application of suction during the anatomical juxtaposition of two *orbicular oris* muscles in a state of contraction'.

For scientific questions, it is important to determine whether the question refers to mainstream science, which is well established and unlikely to change (unless a rare paradigm shift occurs), or the speculative, newer aspects of science, where we are in the process of discovery on the fringes of knowledge. In the first case we should trust the experts for an answer and, unless there is good reason otherwise, we should accept the answer. Thus, the answer to 'did the Universe arise from a big bang 13.7

billion years ago?' is a resounding 'yes' from the vast majority of cosmologists. Or the answer to 'did humans arise by evolution with natural selection?' is also a definite 'yes' from evolutionary biologists. Or again, 'has the Earth's global temperature risen by 1 degree centigrade in the past 100 years?' produces a definitive 'yes' from climate scientists.

For speculative science, we cannot yet trust the answers, which may be either disproved or become part of the mainstream in future. For instance, 'are there many different universes (a multiverse)?' or 'is our 3D space immersed in higher dimensional space?' or 'what happened at or before the big bang?'

Our Universe

A Huge Universe

How much do we know about our Universe? The first fact is that it is almost unbelievably enormous. In the observable universe, there are a hundred thousand million (10^{11}) galaxies, and in each galaxy a hundred thousand million (10^{11}) stars. Thus, in total we have ten thousand billion billion billion (10^{22}) stars.

Fine Tuning

The second major feature is that the Universe is fine-tuned (Rees, 1999), in the sense that extremely small changes in fundamental constants would lead to the impossibility of intelligent life. These constants include

$N = 10^{36}$, the ratio of electric to gravitational forces,
$\varepsilon = 0.007$, the fraction of mass that is converted to energy when four hydrogen atoms combine to form a helium atom,
$\Omega = 0.3$, the density of the Universe,
$\Omega_\lambda = 0.3$, the energy density of a vacuum (the cosmological constant),
$Q = 10^{-5}$, the energy to break up a galactic cluster.

Thus, if, N were increased by 1%, the Sun would explode, and, if $\mathcal{E}$ were 2% bigger, all the hydrogen in the Universe would have been used up a few minutes after the big bang. If the ratio of the mass of a neutron to a proton were very slightly different, then it would not be possible to create carbon, nitrogen, oxygen and so on. The values of Ω, Ω_λ and Q determine the size, age and expansion of the Universe. If they were very slightly different, the Universe would have recollapsed or expanded too rapidly for stars and galaxies to form.

This fine tuning is not a proof of God, but it is consistent with His existence. It leads to the *atheist dilemma,* namely, that either there is purpose in the Universe or there are monstrous coincidences; either there is the working of providence or the Universe is happenstance. For me, the Universe makes more sense if God is behind it, since it gives a coherent understanding for why the universe seems so congenially designed for the existence of intelligent, self-reflective life (Gingerich, 2014).

A third possibility is that our Universe is part of a vaster multiverse of unconnected universes, which are spontaneously created out of nothing, containing different sets of laws of physics. But this suggestion is outside physics since there is apparently no way to prove or disprove it – in other words, it is part of a metaphysical fantasy (Gingerich, 2014). In normal science, a hypothesis has to be testable, but observational evidence for or against a multiverse is most unlikely, so a multiverse is more of a belief than a normal scientific hypothesis.

Indeed, even if a multiverse is somehow confirmed in future, the fine-tuned physical parameters could well still be God-given, and one still needs to understand where the physical laws come from and what 'creation out of nothing' means, since normally creation is a process in time.

The Big Bang

It is now well established that the Universe probably originated in a 'Big Bang', expanding and cooling from a point, 13.7 billion years ago, although I prefer the term 'expanding' or 'unfolding' Universe, since 'big bang' suggests misleadingly a pile of pre-existing explosive that was ignited. The evidence is substantial. Firstly, all galaxies possess a redshift that suggests they are moving away from us at a speed that is proportional to their distance (Hubble's law, 1929). Secondly, the microwave background radiation discovered in 1964 is the glow from the big bang. Thirdly, the helium abundance is well explained as having been formed during the big bang.

Looking at distant objects, we are effectively looking back in time due to the finite time for light to reach us. For example, the Hubble deep field reveals the Universe a few hundred million years after the big bang. But our mainstream understanding goes back much further in time (Fig. 1-10).

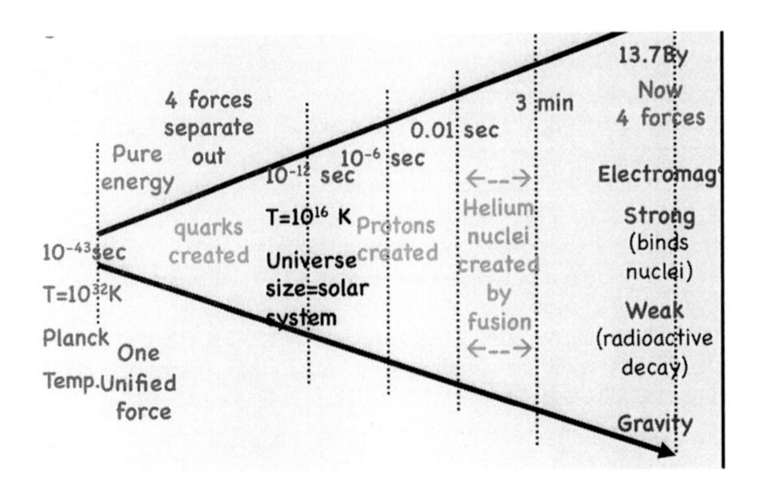

Fig 1-10. Events in the first few minutes after the big bang

Thus, only 10^{-43} sec after the big bang, the temperature was 10^{32} K (the Planck temperature) and there was a state of pure energy with one unified force. By comparison now, after 13.7 billion years, in our Universe there exist four separate forces, namely, the electromagnetic force, the strong force (that binds nuclei), the weak force (responsible for radioactive decay) and the force of gravity.

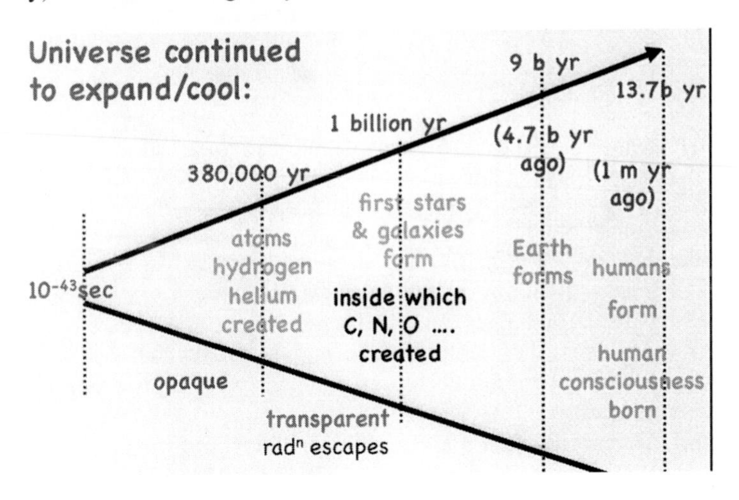

Fig. 1-11. Events later on after the big bang

Very soon after the big bang, the four forces separated out and quarks were created. By 10^{-12} sec, the temperature had fallen to 10^{16} K and the Universe had expanded to fill a volume equal to the present size of the solar system. Then, after 10^{-6} sec, protons were created, and between 0.01 sec and 3 min, all the helium nuclei were created by fusion of protons.

Early in its evolution the Universe was so dense and fully ionized (like the interior of the Sun) that it was opaque, but after 380,000 years atoms of hydrogen and helium were created and so it became transparent as radiation could escape (Fig. 1-11). Also at 380,000 years the microwave background, which we now see glowing at 2.7 K, was created. Indeed, the Planck space mission has observed temperature fluctuations of only 10^{-5} in the microwave background which subsequently grew into galaxies (Fig. 1-12).

Fig. 1-12. Temperature fluctuations in the cosmic microwave background, created from 9 years of data from the Wilkinson Microwave Anisotropy Probe (WMAP) (courtesy NASA/WMAP Science Team)

After 500 million years the first stars and galaxies had formed, inside which the heavier atoms of carbon, nitrogen, oxygen and so on were created by fusion of lighter atoms. Then, after 9 billion years (only 4.7 billion years ago), the Earth formed, and only 200,000 years ago (a twinkling of an eye in cosmic history) humans were formed and human consciousness arose.

Well, all that is part of mainstream cosmology, but what about all the more speculative aspects. The Planck mission has shown us that only 5% of the Universe is made up of normal matter. 27% is so-called *dark matter,* which we can't see but know is there from the observed rotational structure of galaxies, and 68% is *dark energy,* which is associated with the observed accelerating expansion of the Universe. However, we have no idea what is the nature of either dark matter or dark energy!

Dark matter is not normal matter, nor antimatter (which would produce gamma rays when it encounters normal matter), nor black holes (which would produce distorting gravitational lenses). The search is on to identify whether it due to exotic particles such as WIMPs (weakly interacting massive particles).

In the early 1990's, it was expected that the effect of gravity would be to slow down the expansion of the Universe. However, 1998 Hubble Space Telescope observations of distant supernovae showed surprisingly that the expansion was accelerating in time. So-called "dark energy" is thought to solve the puzzle, but its nature is unknown. Several possibilities have been proposed. One is that "empty" space contains its own energy, so that an increase in the size of space creates more energy and so makes the Universe expand faster than it would otherwise do. Another is that dark energy has a negative pressure, which produces a repulsive gravity. This is consistent with a special value of the cosmological constant that Einstein had introduced to keep the Universe static and then abandoned when he found it was expanding.

A further explanation for dark energy is that "empty" space possesses quantum mechanical "virtual particles" that form and disappear and produced a vacuum energy, but its expected value is far too large. Perhaps instead a strange type of energy fluid called "quintessence" fills space.

Another highly speculative part of cosmology in addition to dark matter and dark energy is the question: what happened at extremely early times in the big bang at less than 10^{-36} sec, when quantum mechanics and general relativity break down and need to be unified or combined in some way to give a more general theory.

Two candidates for a new theory of quantum gravity have been proposed. The first is loop quantum gravity, in which space is granular and consists of a network of loops that allow you to consider space before the big bang, which turns out to be instead a "big bounce". The second and most popular candidate is superstring or "M" theory, in which elementary particles are tiny strings rather than points and space has seven extra dimensions that are closed on themselves, crammed into tiny shapes called "Calabi-Yau manifolds" of size only 10^{-33} cm.

In July 2012, the Large Hadron Collider (LHC) in Geneva discovered the Higgs boson, the last particle of the standard model of particle physics. It was expected, however, that dozens of new supersymmetric particles would also by now have been detected, but none have so far been glimpsed. In other words, there is as yet no sign of supersymmetry,

supersymmetry, which is a major failing for the minimal supersymmetric standard model (MSSM). Later this year, the search will be continued at the LHC when it is reopened at higher energies than before. If there are still no detections, the MSSM would have failed and this would be a major crisis for superstring theory, which would seem a far less elegant solution than before.

Thus, the possibilities for our Universe are:

(i) that it is a 3D membrane (or *brane*) cruising through a higher dimensional space, with the big bang being the collision between two branes;

(ii) that the big bang was not the beginning, but was instead part of a big bounce or an oscillating universe;

(iii) that it is part of a multiverse of other universes, spawned inside black holes.

All of these are highly speculative. There is at present no evidence whatsoever for any of them. In future, they may be ruled out, or one of them may become part of the mainstream of understanding.

The Nature of Science

So what is it like to be a scientist? Is it a cold, rational, logical, mechanical existence, undertaken by computers and emotionless people in white coats and having nothing at all to do with arts or Christianity? Also, is the world deterministic, with the weather determined by the behaviour of many individual clouds, life by many individual molecules in cells, and our thoughts by many individual electrical signals in the neurons of our brains?

The first point to make is that metaphysics has a much closer relation to physics than at first thought, with our underlying philosophy of life affecting, often implicitly, our statements about physics. Science and religion use evidence differently but have the same goal of a coherent understanding. Physics cannot be separated from metaphysics. Thus, statements such as 'evolution is atheistic' or 'science tells us we are here by pure chance' are not scientific statements but are coloured by the speaker's personal belief.

The second is that modern science is not clockwork or purely deterministic. It is often a combination of regularity and chance, of law-like and random behaviour. Laws can predict only in general terms, with the appearance of particular stars or sunspots or cyclones being due to statistical or turbulent fluctuations. Regimes of laminar behaviour are interspersed with chaotic behaviour, where two initial conditions that are extremely close can lead to widely different outcomes.

The very nature of space, time and matter are uncertain. Also, there are often several levels of description, with the higher levels having *emergent properties,* which are not predicted from the lower levels but react back down on them. Examples include the weather, the rings of Saturn, flocks of birds, the workings of the human body, and even evolutionary convergence or possibly time itself.

The third point is that being a scientist is much closer in its nature to being a Christian than usually appreciated. In practice, being a scientist involves several features:

(i) Creativity, leaps of faith, intuition and imagination, with moments of inspiration leading to stretches of perspiration as the ideas are worked out using all the skills that have been built up over the years;

(ii) It often fills you with a sense of beauty and wonder and leads to humility as you realize how little you know; if a scientist is being arrogant or closed, he is not being true to the core of science;

(iii) Openness and questioning that lead to a voyage of discovery;

(iv) Trust and integrity, which are essential to being a community

All of these aspects affect my life of faith, and so being a Christian involves just the same features: leaps of faith and trust; a sense of beauty and wonder that leads to humility; openness and questioning, which are my guide on a pilgrimage of faith; and community – the body of Christ. This close similarity suggests an underlying unity between science and Christianity.

Conclusions

My main conclusions are as follows:

(i) The way of the scientist is questioning, open, listening, and affects profoundly my approach to faith;

(ii) The main divisions are not between different denominations or between Christianity, Islam and Atheism, but between those with an open attitude and a closed one; thus, I have more in common with an open-minded humanist than a closed-minded Christian;

(iii) The society that I would like to live in is one where you can discuss openly, listen to and respect those with different beliefs, and where we rejoice in diversity and agree to live at peace;

(iv) For me, a scientist can be a person of faith, provided they are open to the insights of science and responsive to the hand of the Maker in the Universe;

Thus, both the laws of nature and the laws of God are important; as Psalm 19 says

> "The Sun rises from the end of the heavens.....
> The law of the Lord is perfect, reviving the soul".

So, let us enjoy and savour the beauty of creation each day.

References

Barbour, I. (1997) *Religion and Science: Historical and Contemporary Issues,* San Francisco: Harper.

Choi, H. (2004) Presented at *the Science and Religion in Context* Conference, University of Pennsylvania, 2004.

Gingerich, O. (2014) *God's Planet*, Cambridge, MA: Harvard University Press

Gould, S. J. (2002) *Rocks of Ages: Science and Religion in the Fullness of Life*, New York: Ballantyne Books.

McLeish, T. (2014) *Faith and Wisdom in Science*, Oxford: Oxford University Press.

Rees, M. (1999) *Just Six Numbers: The Deep Forces that Shape the Universe*. London: Weidenfeld & Nicolson.

CHAPTER TWO

FROM ANCIENT NATURE WISDOM TO A THEOLOGY OF SCIENCE: A SCIENTIST JOINS JOB'S COMFORTERS

TOM MCLEISH

It is a great delight to be able to contribute to this volume, following the associated conference, on 'Laws of Nature; Laws of God'. As other contributions show, the notion and metaphor of 'law' in science belongs to the modern era, and is contestable how far it can be taken as a faithful description of science. The questions of metaphor and history in science play, in turn, a vital role in our current problematic and tense cultural and political embedding of science itself. I have argued in a recent book (McLeish, 2014) that the reason that it has proved impossible to hold a grown-up conversation in the political sphere on science-related topics (climate change, genetically modified organisms, fracking, *etc.*) is that we have lost a teleological 'cultural narrative' for science itself. We have no answer to the question that asks what science is *for* at a human level, which goes beyond the narrow instrumentalism of technology and economy. Theology is a discipline long experienced in thinking about the idea of purpose, yet its resources have not been sufficiently brought to bear on this issue of human, positive, cultural narratives for science, since this engagement has been framed almost exclusively in the first place by questions of the presumed conflict or complementarity of science and theology. The great benefits that theological thinking can bring to science's greatest needs have been throttled at source.

A contributing habit has been the rather restrictive investigation of Biblical source material within this narrowly-defined science/religion nexus. Perhaps the category of 'law' has itself unwittingly contributed to this: most discussion has focused on creation-texts within the Torah ('the law') in general and within Genesis in particular. The rich seam of Wisdom writings has been far less explored. Within it the notion of 'lawfulness' is balanced by poetic and powerful treatments of what we might term 'lawlessness' – the whirlwind, the earthquake, the storm are at centre stage here, just as much as the regular rising and setting of the sun and the orderly passage of seasons. We might be reminded that our science has also learned to grope its way into a partial understanding of chaotic and random processes. These themes also offer a fresh approach into the human story of intellectual relationship with the material world, within which science constitutes the current chapter.

The Challenges Outlined

So I'd like to persuade readers, alert to the complex interplay of Science and Religion, to look in a few new directions, and to take up some new challenges. In summary:

(1) I suggest that the tense altercations around biblical interpretation of creation narratives are precisely the *wrong* public places to look for where the problems are (they are just the noisy ones, not the important ones). Instead I propose that we re-examine the tense public discourse around science-based technologies, and that the first problem we encounter there is the lack of a narrative understanding of what science is (theologically and anthropologically).

(2) I also suggest that the right place to *be* Biblical in theological thinking about science is not in contemplation of the ornamental, liturgical and geometrical structure of Genesis 1, but listening and debating in the presence of the suffering, pain and disease of Job's ash-heap.

(3) I propose that the right task before us is not reconciliation of Theology *and* Science, but to work through a Theology *of* Science, consistent with the long narrative of creation, fall, and resurrection. My claim is that only this move will restore a consistent 'geometry' to the relationship of science and theology.

(4) I'd like finally to suggest that the carrying out of this task invites a radical reappraisal of what science *is*, culturally, historically and politically, as well as a new model for the 'interaction of science and religion'. The example of environmentalism serves as a vehicle for us.

We will examine each of these ideas in turn before bringing them together in suggesting some consequences for a way forward.

Narratives of Nature

To evidence and explore our first point, I start with nanotechnology – the application of the phenomena of matter at length-scales 10-100 times the atomic, and its special phenomena of self-assembly – biomimetics. Three years ago, a major three-year European research project at Durham University and EU partners explored what was going on behind the ostensibly technical public debate evaluating risks and acceptability. Their project report, *Recovering Responsibility* (Davies *et al.*, 2009) tells a very different story to that of the claims and counter-claims within official reports of public consultations. Its powerful application of qualitative social science unearths underlying narratives of suspicion – stories and themes that influence and permeate the debate, without necessarily surfacing from the superficial technical discussion. As identified by philosopher Jean-Pierre Dupuy (2010), they draw on both ancient and modern myths, and create an undertow to discussion of 'troubled technologies' which, if unrecognised, renders effective public consultation impossible. The research team labelled the narratives:

1. *Be careful what you wish for – the narrative of Desire*
2. *Pandora's Box – the narrative of Evil and Hope*
3. *Messing with Nature – the narrative of the Sacred*
4. *Kept in the Dark – the narrative of Alienation*
5. *The rich get richer and the poor get poorer – the narrative of Exploitation.*

The first three of these Dupuy unites in an 'ancient meta-story', the last two in a 'modern meta-story'. It is at first rather astonishing to find as superficially modern a set of ideas as nanotechnology awakening such a powerful set of ancient stories, but in the light of our claim that the problematic engagement of the human with the material is actually very ancient, and embedded in the discourse of sacred texts and the stories of their communities, it becomes less so. Surveying briefly how they play out:

New technologies, especially those whose functions are hidden away at the invisible molecular scale, promise much, and have made exaggerated claims of benefits: longer, healthier lives at low cost, self-repairing materials and machines, built-in sources of energy. But such hubris elicits memories of over-promising – so, "be careful what you wish for".

The Story of Pandora's Box enters at this point, for this tale of the seductive power of the hidden speaks across the ages to our power to unlock the twinned histories of trouble and hope. The nanotechnological study identifies irreversibility in both knowledge gained and in the 'release into the environment' of nanoparticles. Pandora also released hope from her casket – in the original myth usually read as a positive and counteracting good. However, as Dupuy points out, hope can be dangerous: it can drive a course of action onwards beyond the point at which a dispassionate risk analysis would have recommended a halt.

The third 'ancient narrative' is a fascinating and perplexing one. Why would a secular age develop a storyline that warns us away from 'Messing with Nature' because of its sacred qualities? The secularisation of thought and society has been charted, in the last century, in social theory from Emil Durkheim and in political philosophy from Hannah Arendt. Even the more recent social analyses of the persistence of religious thought into the modern world, such as that of Jürgen Habermas (2006), have approached religions as minority communities. But "the sacred" persists both within and without official religious communities.

Here is a fascinating example, this time in what is surely an analogy to another troubled technology, the process of fracture-recovery of coal-gas from near-surface shales known as 'fracking':

> In ancient times, people believed that inclement weather came directly from a divine source: Whether it be Gods, Goddesses, or just the "spirit of the planet", we have always arranged sacrificial offerings and desperately tried to appease whichever deity has punished us for our wickedness. Although we have somewhat "grown out of" this concept of divine retribution for sin, we kind of have to admit that we have become sinful in our collective attempts to thwart nature and impose our will upon it (Geo & Geo, 2015).

The fourth narrative of being "kept in the dark" is at first sight, as Dupuy observes, a more modern one, speaking of asymmetries in political power between the governing and the governed.

The fifth narrative of "the rich get richer and the poor get poorer" extends the fourth: with exclusion comes, at least, lack of access to the benefits of knowledge and, worse, unequal exposure to their harmful consequences. This fifth narrative has, for example, been especially prevalent in the resistance to GM crops in India.

The European Nanotechnology study is interesting, not only because it enables us to begin to make progress in perceiving why our newest technologies are so troubled, but also through its unearthing of the fundamental importance of underlying narrative: here there are (at least) five ancient narratives coiling around a resistance to new science and new technology. They highlight in the most lurid possible contrast that science itself has no such source to draw on – *there is a narrative vacuum where the story of science in human relationship with nature needs to be told.* What might happen to public debate on contentious science and technology if there were an active ancient narrative that was more neutral, or even positive, in its recounting of our exploration of nature? I have elsewhere (McLeish, *loc. cit.*) suggested that the late-modern label of "science" may itself be part of the problem. Freighted as "science" is with the etymology of a knowledge claim, the adoption of the earlier term "natural philosophy" carries instead a humbler and warmer set of connotations – a "love of wisdom to do with natural things".

Is there any modern articulation of an underlying narrative structure to science? Finding a clue in one example from the academic world, George Steiner writes this about art in his deeply felt discussion of meaning and language, *Real Presences*: "Only art can go some way towards making accessible, towards waking into some measure of communicability, the sheer inhuman otherness of matter…". "Only *art*?" To a scientist this comes as a shock – to me as a wake-up call to think what really motivates us, at the deepest level, to explore the world. If science is not there to establish lines of communication between our minds and the "sheer inhuman otherness of matter", then what is it doing? Why does Steiner, so sensitive to our human need for some sort of reconciliation with our world, not see science as part of the answer when it surprises us over and over again with our ability to reveal the patterns beneath the things we see, hear, feel and touch by careful observation, imagination and theory? Perhaps in spite of his apparent familiarity with the parade of the science community, he speaks for many people when he denies it a role in the inspired contemplation and recreation that he does see in art.

We might note at this point, that none of Dupuy's ancient or modern narrative lines detected in the public framing of nanotechnology, nor Steiner's felt need to negotiate a gulf between the human and the material, speak in the category of "law". The nearest we come is the fabled pattern of unintended consequences in the Pandora story, but this is more reminiscent of common wisdom than even natural, let alone scientific, law.

By Job's Ash Heap

I have often suggested to my scientific colleagues that they pick up and read through the closing chapters of the Old Testament wisdom book of Job; they later return with responses of astonishment and delight. I did the same on my first reading. Perhaps this is because of its celebration of the chaotic and the unknown, perhaps because of its repeated use of the question-form – surely the imaginative well-spring at the core of science itself? Let us taste some of its beauty right away, from the point at which God finally speaks to Job (after 37 chapters of silence!) in chapter 38, v4[1]:

Where were you when I founded the earth?
Tell me, if you have insight.
Who fixed its dimensions? Surely you know!
Who stretched the measuring cord across it?
Into what were its bases sunk,
or who set its capstone, when the stars of the morning rejoiced together,
and all the sons of God shouted for joy?..

We are familiar with this type of language; it is a beautiful development of the core creation narrative in Hebrew wisdom poetry (a form found in Psalms, Proverbs and some Prophets too, that speaks of creation through ordering, bounding and setting foundations), but now in the relentless urgency of the question-form, the voice continues:

Where is the realm of the dwelling of light, and as for darkness, where is its place?

So from the creative process of ordering, bounding and shaping it asks about the fundamental form of light, then sharpens its questions towards the phenomena of the atmosphere:

Have you entered the storehouses of the snow?
Or have you seen the arsenals of the hail ...
Where is the realm where heat is created, which the sirocco spreads across the earth?
Who cuts a channel for the torrent of rain, a path for the thunderbolt?

The voice then directs our gaze upwards to the stars in their constellations, to their motion, and to the laws that govern them:

Can you bind the cluster of the Pleiades, or loose Orion's belt?
Can you bring out Mazzaroth in its season, or guide Aldebaran with its train?
Do you determine the laws of the heaven?
Can you establish its rule upon earth?

The questing survey next sweeps over the animal kingdom:

---

[1] All Biblical quotations in this chapter are from the magisterial new translation and commentary by David Clines (2011, Vol. 3).

Do you hunt prey for the lion, do you satisfy the appetite of its cubs,
while they crouch in their dens, lie in their lairs in the thickets?

and at the glory of flight in both its migratory navigational intelligence and mastery of the air:

Is it by your understanding that the hawk takes flight, and spreads its wings toward the south?
Is it at your command that the eagle soars and makes its nest on high?

It finishes with the celebrated "de-centralising" text that places humans at the periphery of the world, looking on in wonder at its centre-pieces, the great beasts Behemoth and Leviathan:

Beneath it are the sharpest of potsherds; it leaves a mark like a threshing sledge upon the mud.
It makes the deep boil like a cauldron; it makes the water [bubble] like an ointment-pot.
Behind, it leaves a shining wake; one would think the deep hoar-headed!
Upon earth there is not its like, a creature born to know no fear.
All that are lofty fear it; it is king over all proud beasts.

Where is this voice coming from, that resonates with question after question? The answer is itself a fascinating surprise. At the very start of this passage, known as "The Lord's Answer" we are told:

And Yahweh answered Job from the tempest

so situating the entire monologue within one of the wisdom tradition's great metaphors for chaos. Commentators have been quick to note that none of the animals appearing in the poem is domestic, nor are any of the cosmic powers, or forces it asks about, controlled by humans. This is an ancient recognition of the unpredictable aspects of the world: the whirlwind, the earthquake, the flood.

Even these short extracts from the much longer poem give something of the impressive, cosmic sweep of this text, the grandeur of its scope, and the urgent, pressing tone with which it peers into the nooks and crannies of creation. In today's terms, we have in the Lord's answer to Job as good a foundational framing as any for the primary questions of the fields we now call cosmology, geology, meteorology, astronomy, zoology … Of course to use the text in that way is an unwarranted and anachronistic projection of our current taxonomies and programs onto a quite different genre of literature and over a vast gulf of cultures. However, if we are instead alert to the poetic form, we can recognise in this extraordinary wisdom-poem an ancient and questioning view into nature, unsurpassed in its astute attention to detail and sensibility towards the tensions of humanity in confrontation with nature. There are forces at play behind this text that lie at a depth and draw on an energy that still lie at the roots of the relationship between the human and the non-human worlds. The public projection of science today is still unfairly dominated by a deterministic Newtonian (or quantum for that matter) paradigm. This is also true of material recruited into most science-religion debates. But a scientist alert to the far more subtle ergodic dynamics underpinning the many-body physics of statistical mechanics, or of the chaotic phenomena in nonlinear dynamical systems theory, will find herself at home in Job's world of incomprehension in the face of cosmic disorder.

As well as its universal and cosmic content, there is another reason that scientists find this passage in Job so resonant – and that is its *form*. For we know that the truly essential and imaginative task in scientific discovery is not the finding of answers, but the formulation of the fruitful question. Werner Heisenberg, the pioneer of quantum physics, put it this way during his 1955 Gifford Lectures at St Andrew's University (Heisenberg, 1957): "What we observe is not nature itself, but nature exposed to our method of questioning."

Long recognised as a masterpiece of ancient literature, the Book of Job has attracted and perplexed scholars in equal measures for centuries, and is still a vibrant field of study right up to the present day. David Clines, to whom we owe the translation employed here, reproduced from his recent edition and commentary, calls the Book of Job "the most intense book theologically and intellectually of the Old Testament". It is intriguing that, ubiquitous in biblical nature-writing, ideas about the created world are woven into a text that takes pain and suffering for its theme.

However, although readers of the text have long recognised that the cosmological motif within Job is striking and important, it has not received as much comprehensive attention as the legal, moral and theological strands in the book. This de-emphasising of cosmology might partly explain why the long passage from which we have taken the extracts above, known as "The Lord's Answer", has had such a problematic history of reception and interpretation. Does it really answer Job's two questions about his own innocence and the meaninglessness of his suffering? Does the "Lord" of the creation hymns correspond to the creator Yahweh of the Psalms, the Pentateuch and the Prophets? Does the text even belong to the rest of the book as originally conceived? Some scholars have found the

Lord's answer to Job spiteful, a petulant put-down that misses the point and avoids the tough questions. Others, partly in sympathy with that interpretation, have suggested that the entire discourse has been "glued on" to the earlier chapters at a later date and by a different author, pointing out that a simple contrast of God's knowledge of nature to Job's ignorance admits of no apparent satisfaction to his complaints. So, for example, Robertson (1973) perceives that this "God" fails utterly to answer Job, finding him a charlatan deity. Even those who take a very different view find the Lords Answer presenting an over-tidy view of the world, so Clines claims of it, "There is no problem with the world. Yahweh does not attempt a justification for anything that happens in the world, and there is nothing that he needs to set right. The world is as he designed it."

But are these interpretations justified? Increasingly scholars have recognised an underlying unity to the book that makes it hard to escape the tough questions around the Lord's Answer. Even looking at the text through the fresh lens of science today (even if that lens is at the wrong end of a telescope) resonates with the *difficulty* of questioning nature, even its painfulness, as well as its *wonder* – that is how scientists respond at a first reading time and again.

To bring these threads together one can take a journey through the Book of Job, travelling on the ground of a "close reading", taking one path, albeit a not so well-trodden one, to the snowy peak where the Voice from the Whirlwind speaks. For there is a track through the book that starts with the workings and structure of the natural world, and, while winding through the arguments of the disputations, never leaves it. This is the reading natural to the scientist and, on its pathway, one thing we notice is that all the natural images invoked in The Lord's Answer have actually already appeared in the disputation speeches of Job and his friends, from the great beast Leviathan to the roots of the trees and the birds of the air, from the desert floods filling the wadis below, to the starry heavens above. A structure that helps us summarise what such a "nature" trail through Job looks like[2] recognises a distinct pattern in the realms of creation, explored predominantly in the three cycles of speeches between Job and his three companions:

The First Cycle takes its imagery from inanimate creation, so we find references to earth, winds, waters, springs, stones, sea, *e.g.* Eliphaz' superlative terms in which to frame the rewards of Job's repentance (5v.22):

---

[2] A fuller version can be found in McLeish (2014), chapter 5.

At ruin and blight you will mock, and you will have no fear of the wild beasts.
For you will be in covenant with the stones of the field, and the wild animals will be at peace with you.

The Second cycle moves to a new level of complexity in living things, both plants and animals and their consumable products. So we find vines, cattle, milk, honey: *e.g.* Job's final speech of the cycle evidencing the absence of moral code with nature:

Their bull sires without fail, their cow gives birth and does not lose her calf...
How often are they like straw before the wind, like chaff swept away by the storm?
His pails are full of milk, and the marrow is juicy in his bones.

The animate, inanimate and human fields of the physical world all declare with Job that there is no correlation between morality and matter.

The Third Cycle represents the climax of the increasingly tense and accusative argument between Job and the companions. For the first time the nature-imagery reaches the heavens, moon, stars, Sheol, and the far extremities of the world. Thus Eliphaz introduces us to our first explicit view of a Hebrew cosmology (22v12-14):

Is not God in the height of the heavens? Does he not look down on the topmost stars, high as they are?
Yet you say, "What does God know? Can he see through thick clouds to govern?
Thick clouds veil him, and he cannot see as he goes his way on the vault of heaven!"

So the cosmological crescendo matches the ratcheting tension of the drama – it is only in this third cycle that direct accusations of personal wickedness are levelled at Job, a brutal climax where Bildad can only answer Job's searching complaints by an attempt to trump them with their irrelevance. God, all-powerful, may rule with an iron fist if he so desires. Finally there is no human voice worth hearing in such a world: all voices are crushed into silence.

A composer has a hard job when the piece of music being shaped reaches an ugly and deafening climax. Where do you go when the full forces of the orchestra have once joined their different themes together, taking the listeners to a moment of overwhelming excitement, but also of terror? Perhaps this is one reason why the later editors of the text get confused at this point, why some versions assign no third-cycle speech to Zophar at all, and why Job's responses become snatched and stylised. It happens in music too: themes disperse, shattered into different voices, attempting to regroup, to find an answer to the experience they have just lived through (think of the aftermath to the thunderstorm in Beethoven's *Pastoral Symphony*, or the dawning of day in Mussorgsky's *Night on a Bare Mountain*). The alternative, as adopted by Ravel in *Bolero*, is of course just to end the whole thing right there. Some commentators on Job have suggested exactly that of the original text. But that is not what happens in the version we have received from antiquity. More Mussorgsky than Ravel, the disordered and frightening impasse is broken into by a new voice with a new subject. Or perhaps it is really an old subject, deeply buried but now resurfacing. Let's listen to the start of chapter 28, a beautiful and structurally quite new voice sometimes denoted "the hymn to wisdom" (28v1-6):

> *Surely there is a mine for silver, and a place where gold is refined.*
> *Iron is taken from the soil, rock that will be poured out as copper.*
> *An end is put to darkness, and to the furthest bound they seek the ore in gloom and deep darkness.*
> *A foreign race cuts the shafts; forgotten by travelers, far away from humans they dangle and sway.*
> *That earth from which food comes forth is underneath changed as if by fire.*
> *Its rocks are the source of lapis, with its flecks of gold.*

The scene is a mineshaft under the ground, and the voice is a miners' song! Foreign workers in the ancient middle east were commonly employed in such dangerous occupations; here we picture the mineworkers tunnelling and cutting the rock. Roped to the subterranean rock-face, we can just make them out swaying in the gloom. We also begin to see *with* them: a miner's gaze on the earth from below reveals a very different appearance to that from above. The "transformation as if by fire" is a remarkable insight into one of the processes by which minerals separate out, recombine and solidify in the rocks below ground. If we look hard in the dim candlelight we might catch a glimmer of gold. The discourse of earlier chapters touched in one or two places on smelting and refining metals from ores, but this song takes us much further back into the process of extracting the ores, and in an entirely new setting. It even begins to probe how the ores might have arrived there. The underground world takes us completely by surprise – why did either an original author or a later compiler suppose that the next step to take in the book was down a mineshaft? Reading on,

There is a path no bird of prey knows, unseen by the eye of falcons.
The proud beasts have not trodden it, no lion has prowled it.
The men set their hands against the flinty rock, and overturn mountains at their roots.
They split open channels in the rocks, and their eye lights on any precious object.
They explore the sources of rivers, bringing to light what has been hidden.

We begin to recognise a tune that has been with us all along our nature trail through Job; namely that there is something especially human about the way we fashion our relationship to the physical world. It affects where we go (*There is a path*), what we see (*unseen by falcons*), what we understand (*bringing to light what has been hidden*) and what we do (*split open channels in the rock*). This extraordinary power to connect with nature seems so strongly worded that some readers have assumed that here the song is really talking about God, the creator himself, not humans at all. To take a specific example, "overturning mountains at their roots", sounds like the exercise of divine power, but if we bear in mind that the Hebrew word (*haphak*), translated "overturning", is just the same as that for "changing" (as in "changed by fire"), the metaphor directs us rather to admire the patient and knowledgeable art of mining seams through hard rock, exploring just those places that yield precious ores or stones. But, even more significantly, only human eyes can *see* it from this new viewpoint. It is a sight that asks questions, that directs further exploration, that wonders. No wonder such extraordinary human capacity has been confused with the divine.

Until now, the writer has kept to himself the primary subject of the hymn in chapter 28, which will answer our question of what brought us to this hidden place. But now we are let into the secret, and a new question sounds (28v12):

But where is wisdom to be found? And where is the place of understanding?
Humans do not know the way to it; it is not found in the land of the living.
The ocean deep says, "It is not in me," and the sea, "Not with me."

We have been on a quest for wisdom and understanding all along! These are two different Hebrew words as they are in English, meaning two different things: the first a general idea of practical knowledge, the second a more intellectual and contemplative grasp. Carol Newsome, in her commentary on Job, points out that their juxtaposition can signify something more, "the kind of understanding that would provide insight into the nature and meaning of the entire cosmos". And a timely search it is – surely wisdom is what the court of appeal will need to resolve Job's demands and the friends' unsatisfactory replies. Perhaps the idea that the type of wisdom we will need is the same that searches the workings of the cosmos does not seem quite so strange, after our journey through the circling arguments themselves, and their continual return to the natural world, as it might have previously.

The writer tells us that wisdom is hidden from the eyes of all living things (humans, presumably, included), and that even the deeply buried land of death has only heard enigmatic whispers of it. So is the world simply a collection of "the wrong places to look" for wisdom and understanding? The conclusion of the hymn has driven different readers to opposing views, by "drawing back the curtain" once more (28v23):

But God understands the way to it; it is he who knows its place.
For he looked to the ends of the earth, and beheld everything under the heavens,
So as to assign a weight to the wind, and determine the waters by measure,
when he made a decree for the rain and a path for the thunderbolt –
then he saw and appraised it, established it and fathomed it.
And he said to humankind,
Behold, wisdom is to fear the Lord, understanding is to shun evil.

The answer is for some commentators a deep disappointment, even a banality (as Clines) – is it really true that, after all this exploration of nature from above and below in a search for wisdom, all that is futile and one has only recourse to a pious "fear of the Lord"? For an imaginary humanity without history, or rather without an intellectual future that contains more than the present, perhaps this is so. But it is by no means true that the wisdom hymn concludes that wisdom has nothing to do with the created world, for the *reason* that God knows where to find it is precisely because he "looked to the ends of the earth, …, established it and fathomed it"! It is, as for the underground miners, a very special sort of looking – involving number (in an impressive leap of the imagination in which we assign a value to the force of the wind), physical law (in the controlled paths of rain and lightning) and formation (there is a blurring here between creating the world, and looking at it once it is created). This is an extraordinary claim: that wisdom is to be found in participating with a deep understanding of the world, its structure and dynamics. It is banal and disappointing only if the very final injunction to humans is to be taken as an instruction to turn from numbering, weighing and participating in nature as God's exclusive domain, and restrict our thoughts and behaviours to the moral sphere alone. Is the writer really saying that that sort of "fear of the Lord" is the end of all wisdom? I'm not sure we really listen properly to the latent meaning of the word "*beginning*" in the closely parallel, and well known, opening to the Book of Proverbs, "The fear of the Lord is the beginning of wisdom" (Pr 1v7). Wisdom, in Proverbs, is not a state but a path – even, of course, a person – (the writer of Job would employ what is becoming a favourite word: *derek* – "a way"). We know already that the momentum of the Book will not remain circular, nor will God always remain speechless, but will extend to Job, and to the readers of the Book, an extraordinary invitation to engage with him, and especially with his "fathoming" of the universe. If the "fear of the Lord" carries a higher meaning of engagement, following and exploration, rather than a simple moral obeisance, then the end of the hymn to wisdom is far from a banal journey's end. Instead it becomes a signpost to a mountain top view only currently obscured from us.

Looking back at the three cycles of dialogue with this new perspective, we see not only that they have taken us through ascending levels of the natural world itself, but that they have also introduced us to a series of different interpretations of what a human *relationship* with nature might look like. Getting this relationship right, we now understand from the cornerstone passage of the wisdom hymn, offers us the prospect of a route to the precious possession of wisdom, the quality most starkly lacking from the disputations. However, none of the candidate relational perspectives offered from the clamouring voices within the Book of Job has succeeded. At least five can be discerned beneath and between the voices we hear by the time Elihu finishes speaking. It's worth summarising them.

First is the "simple moral pendulum" – the story of nature as both anthropocentric and driven by a moral law of retribution. This is the central narrative of the first three of Job's friends – their underlying simplistic, deterministic (and ultimately barbaric) worldview is effectively unchanged throughout the conversation.

Second is the "eternal mystery" – the story that speaks of God's exclusive understanding of nature's workings in ways that humans can never know. It picks up illustrative weight in the repetitive use of untamed animals, distant points of the compass, and cosmological structures far removed from us in time and space.

Third is the "book of nature" idea – the story in which nature constitutes a giant message-board from its maker, for those who have eyes to read it. Attaining its height in Elihu's speech, humans are central to this relationship just as are pupils in a classroom, but this classroom belongs in a kindergarten not a university.

Fourth is the story of the uncontrolled storm, flood and earthquake. This is uniquely Job's interpretation of his relationship with nature, but extrapolated in his anguish and exasperation. Through this lens, creation is chaotic rather than regulated, and bound over to a crumbling decay. Humanity is swept up in the storm and flood, which God might have held at bay, but chooses not to. *Job uses cosmological chaos in his discourse because his accusation is that God is as out of control of human justice as he is of the physical world.*

A fifth possible relationship with creation is made explicit only once, by Job himself. It is the relationship of nature-worship. It is dismissed straight away, but not without giving away its allure.

A sixth storyline is hinted at, but not spelled out with clarity. It has something to do with the centrality of the created physical world over any claim by humanity to a pivotal place within it, yet it is the voice that locates wisdom within perspicacious knowledge of nature. It hints at a balance between order and chaos rather than a domination of either. It inspires bold ideas such as a covenant between humans and the stones, thinks through the provenance of rainclouds, observes the structure of the mountains from below, wonders at the weightless suspension of the earth itself.

This is the point at which we first entered the Book of Job in this essay, the point at which David Clines admits (surely self-referentially) a "frisson within" on every reading, "even for those who have grown old with the Book of Job". Yahweh finally speaks, at the very point where everyone else has finally fallen silent. He plucks Job up from his ash-heap and takes him, and us, on a whirlwind tour of creation from its beginning to the present, from the depths to the heights and from its grandest displays into its inner workings.

But one thundering question remains for all readers of the book: does Job receive an adequate answer to his two complaints, in the Lord's Answer? But now that we have seen that the question of God's justice in his management of creation as a whole is woven into Job's disputations, we can see that it is not by-passing the question for the Lord's answer to take this thread and expand it into the glorious quest into nature's workings with which the book finishes (or nearly finishes). With trepidation, and against the weight of opinion, I am therefore suggesting that the 'Lord's Answer' *is* an answer to Job's complaint, possibly the only adequate answer.

Firstly it tackles head-on the accusation that creation is out of control, by suggesting ways of thinking through what Job's (and his friends') idea of "control" might mean. The deterministic and predictable response of a cosmos that metes out retribution on the unjust is not a living universe but a dead one. The axis of control and chaos is subverted by the revelation of a third path of constrained freedom, in which true exploration of possibility, of life, really lies.

Secondly it does, against all expectation, achieve what has always been Job's aim, to be reconciled to his state of physical pain and mental outrage. We are not privy to Job's inner response to the Lord's answer, we don't know the steps that lead him to aver at the end (42v5)

I have heard you with my ears, and my eyes have now seen you.
So I submit, and I accept consolation for my dust and ashes.

But we do know that he has been lead towards a radically new perspective, one that in one way totally de-centralises humanity from any claim to primacy within creation, yet in another affirms the human possibility to perceive and know creation with an insight that is at least an image of the divine one.

Thirdly it is participative and invitational: the final Voice asks the great questions about nature not, purely, to rouse us into self-awareness of our own lack of understanding, but as an invitation towards transforming it in encounter with wisdom.

Fourthly it speaks of the fundamental significance and importance of the physical structure and workings of nature. They are not sideshow or an optional hobby for the socially challenged. Our relation of perception, knowledge and understanding is at the centre of our humanity. Job even talks of the direction of human relationship with creation in terms of a covenant.

Fifthly the Lord's Answer is "eschatological" – its message is one that announces and urges the possibility of a future in which this vital relationship, now broken, becomes healed, not just for Job but for the species from which he comes. Job may not know the answers to Yahweh's questions, but one day he or his descendants might well do[3].

---

[3] This suggestion has been made at least once before, by David Wolfers in his brief "Science in the Book of Job", *Jewish Bible Quarterly*, 19, 18-21 (1990). His conclusion runs, "The majority of these questions [those of scientific purview] are to be found in the Lord's first speech to Job, and there is little doubt that their primary purpose is to expose the abysmal ignorance of mankind of all theoretical

The eschatological drive of the text constitutes another pointer – that within these wisdom writings is the deep theme of purpose, a recognition that all human action belongs within a linear history between creation and re-creation. This is the basis on which we can draw on the sixth 'storyline' in Job, for material we need to construct a sixth narrative for science in our own times - one that Du Puy's analysis reveals as occupied by a narrative vacuum. The next step is to derive a 'theology of science' – an account of what doing science signifies and achieves within covenant history and its New Testament articulation as "working out salvation" (Phillipians 1v12).

A Theology of Science

We need to draw together some threads from readings of Job and other wisdom texts, considered alongside our experience of doing science. Most of the constitutive themes for a theology of science that I extract in detail in *Faith and Wisdom in Science* have already emerged in our examination of Job. They are:

- long and linear history of engagement with nature,
- surprising human aptitude for reimagining nature,
- search for wisdom as well as knowledge,
- ambiguity and experience of pain,
- delicate balance of order and chaos,
- centrality of the question and the questioning mind,
- experience of love.

These are the lines that draw us to a larger narrative in which science can be framed.

Within all these themes the pattern of relationship has dogged us constantly. Science experiences the negotiation of a new relationship between human minds and the physical world. The nature-language of the Bible is consistently employed to describe and develop the relationship of care and of understanding between us humans and a world that is both our home and also a frightening field of bewildering complexity. Although fraught with ambiguity, experiencing pain and joy in equal measure, knowing terror before the phenomenon of chaos as well as experiencing joy before its resplendent order, bewildered by ignorance yet granted hard-

---

aspects of Creation. Is it possible to detect a faint hint, that it might be well for man to set about attempting to remedy this ignorance?"

won understanding, the Biblical theology of nature is consistently relational.

St Paul invested to a deeply personal degree in the nascent Christian communities with which he worked. Within his most painful correspondence (with Corinth) he re-thinks the entire project of God's creation in relational terms, working around and towards the central idea of reconciliation. The argument begins with the fifth chapter of his second letter to the Corinthians, recapitulating briefly his picture of a "groaning" creation, from the letter to the Romans, in longing for a more eternal form, which he calls *clothed with our heavenly dwelling, so that what is mortal may be swallowed up by life*. Arguing that those who have been baptised into the life with Christ can already view the world from the perspective of its future physical re-creation, he writes (2Cor v17):

> *Therefore, if anyone is in Christ – new creation;*
> *The old has gone, the new has come!*
> *All this is from God, who reconciled himself through Christ and gave us the ministry of reconciliation:*
> *That God was reconciling the world to himself in Christ*

The *ministry of reconciliation* is a stunningly brief encapsulation of the Biblical story of the purpose to which God calls people. I don't know a better three-word definition of Christianity, and it does very well as an entry point for Old Testament temple-based Judaism as well.

There is one relationship that tends to be overlooked in expositions of Christian theology – perhaps humbler than the more obviously broken human ones, but just as profound. It is the relationship between humankind and nature itself. A theology of science, consistent with the stories we have told up to this point, situates our exploration of nature within that greater task. Science becomes, within a Christian theology, the grounded outworking of the 'ministry of reconciliation' between humankind and the world. Far from being a task that threatens to derail the narrative of salvation, it actually participates within it. Science is the name we now give to the deeply human, theological task and ancient story of participating in the mending of our relationship with nature.

It is an extraordinary idea at first, especially if we have been used to negotiating ground between 'science' and 'religion; as if there were a disputed frontier requiring some sort of disciplinary peacekeeping force to hold the line. It also makes little sense within a view of history that sees science as an exclusively modern and secular development, replacing outworn cultural practices of ignorance and dogmatic authoritarianism with 'scientific method' and evidence-based logic. But neither of these assumptions stands up to disciplinary analysis on the one hand, or to an informed view of history on the other.

We might also note that such a participative and active theological understanding of science stands in radical opposition to the tradition of "natural theology". Most fully developed in the eight 19th century "Bridgewater Treatises" this tradition (Robson, 1990; also Chap 4 herein), and in the earlier words of 17th century naturalist John Ray, seeks "to illustrate the glory of God in the knowledge of the works of nature or creation" (Armstrong, 2000). But, rather than looking into nature and seeing God, science as a participatory ministry of reconciliation stands *with* God (in the relation of Job 28's Hymn to Wisdom) looking with informed vision *in imagio Dei* and a concerned creative love. The human-divine axis differs fully 180° from that of natural theology.

Neither science nor theology can be self-authentic unless they can be universal. We need a "theology of science" because we need a theology of everything. If we fail, then we have a theology of nothing. Such a theology has to bear in mind the tension that the same is true for science – it has never worked to claim that science can speak of some, but not of other topics. Science and theology are not complementary, they are not in combat, neither are they merely consistent – they are "*of* each other". This is the first ingredient of a theology of science.

Just as there is no boundary to be drawn across the domain of subject, there is no boundary within time that demarcates successive reigns of theology and science. It is just not possible to define a moment in the history of thought that marks a temporal boundary between the "pre-scientific" and "scientific". The questioning longing to understand, to go beneath the superficialities of the world in thought, to reconstruct the workings of the universe in our minds, is a cultural activity as old as any other. Furthermore, it is a human endeavour deeply and continually rooted in theological tradition. The conclusion is still surprising: far from being necessarily contradictory or threatening to a religious worldview in general, or to Christianity in particular, science turns out to be an intensely theological activity. When we do science, we participate in the healing work of the Creator. When we understand a little more of nature, we take a step further in the reconciliation of a broken relationship.

Consequences for Science in Culture: the example of environmentalism

Does a theology of science do meaningful work for us? Does it provide any avenues to resolve the painful cross-currents around science in society?

Does it suggest new tasks? These must be the test for any endeavour of this kind. Does such an outline for a theology *of* science begin to circumvent a relentless territorial contest between theology *and* science? Let's look at just one example;

One leading contemporary commentator whose interest in the "politics of nature" has not been marginalized is the French thinker Bruno Latour (2004). In a recent edited volume he explores the terrifying observation that "environmentalism" has become a dull topic – with conclusions that are remarkably resonant with our own. They break down into four findings, in his own words: *a stifling belief in the existence of a Nature to be protected; a particular conception of Science; a limited gamut of emotions in politics; and finally the direction these give to the arrow of time.* This is a grand, overarching critique of the politics of nature, but even so, it homes on to the same narrative analysis as did the specific nanotechnology study we examined at the beginning. Latour's identification of the "stifling" move to withdraw all human corruption from a "Nature" that should be maintained in some pristine condition, is none other than the "messing with sacred Nature" narrative by another name. Latour extracts the self-contradictory structure of this story of the Golden Age – Nature reserves are artificial by definition. But the alternative "modernist" trajectory is no less problematic. There the story is an overcoming of Nature with control. We disengage from our environment, not through an "environmentalist" dream of withdrawal from the sanctuary, but through technological domination. Here Latour revisits the narrative of Pandora's Box, because such a modernist hope is dashed on the rocks of the same increasingly deep and problematic entangling with the world that prevents withdrawal. Nature does not respond mildly to an attempt to control or dominate. So neither narrative works – both start with fundamentally misguided notions of the geometries and constraints of our relationship with nature.

Latour's critique of the conception of science is equally resonant with the flawed view of the notion of a "scientific priesthood". Political action on scientific decisions is as paralysed by disagreement as it is by disengagement. Not every expert agrees that blood transfusion might transmit the AIDS virus – so we wait in inaction that condemns children to infection. There is no uniform view on the future trajectory of global warming and its connection with human release of carbon dioxide – so we meet and talk, but do not implement. This is the "kept in the dark" narrative with a twist – the political and public community self-imposes ignorance by demanding that scientists behave as a conclave, reading the same script and praying the same prayers, until the white smoke of expert agreement is released. The political life-blood of a communally-possessed and confident debate, widely shared and energised, respecting where specialist knowledge lies but challenged within a participating lay public, is simply not yet flowing in our national and international veins.

At the close of his contribution to *Postenvironmentalism*, Latour makes an extraordinary move – one that meets our own journey head on. He calls for a re-examination of the connection between mastery, technology and *theology* as a route out of the environmental impasse. We have not yet remarked that the ancient narratives unearthed by the nanotechnology project, and reflected in Latour's, are all implicitly or explicitly pagan, though we have seen how they might be met with, and transformed by, the more positive themes of a Judeo-Christian "ancient narrative" of nature. So when he refers to the Christological theme of the creator who takes the responsibility to engage with even an errant creation to the point of crucifixion, the contrast with the disempowering and risk-averse narratives of "being careful what you wish for", *Pandora*, sacred nature, and the rest, could not be starker.

The theological wisdom tradition we have been following, especially in the way that it entangles with the story of science itself, has brought us to the same point that Latour reaches from the perspective of political philosophy. One identifies the need, the other the motivation and resource, for a re-engagement with the material world, and an acknowledgement that one unavoidable consequence of being human is that we have, in the terms of the Book of Job, a "covenant with the rocks". This extraordinarily powerful collision of metaphors surely points to the balanced and responsible sense of "mastery" that Latour urges that we differentiate from the overtones of exploitative dominance.

More is true – if we take one by one the strands of the "theology of science" that we teased out of our Biblical nature trail, it begins to look as though they might be woven into the story, the missing narrative, that Latour wants to hear told. It is a radical narrative, too, far from the conservatively visceral reaction to environmental technologies that cries out that we must never "meddle with nature". Instead, a Christian voice can hope to take a measured sense of responsible adventure into a more complex but much more authentic and obedient outworking of a commission to exercise wise, not exploitative, dominion over the natural world. It is not by any means obvious, for example, that genetically modified crops should not be developed for pest-resistance or added nutrition. But a wise regulatory process will balance their potential for creative good with an assessment of harms that they potentially unleash if harnessed primarily to a profit motive (McNaghten & Carro-Ripalda,

2015). Such a process will engage in a radical way the welcome new agenda of "*Responsible Research and Innovation*" (RRI: Owen *et al.*, 2012), regarding public communication of research and technology not as a "marketing exercise" towards the end of a development programme, but an essential part of anticipatory and ethical evaluation of both process and outcomes. This richer public conversation now allows space for the recognition of the unknown and the uncertain, the need to take risks, and the need to establish contingent alternatives.

In Latour's words once more, such a pattern for environmental science and innovation will extract its "idea of mastery and creation from God; well the Christian God at least is not a master who masters anything (in the first modernist sense of the word) but who, on the contrary, gets folded into, involved with, implicated with and incarnated into His Creation; and who is so much attached and dependent on His Creation that he is continually forced (convinced? willing?) to save it again and again."

References

Arendt, H. (1958) *The Human Condition*. Chicago: University of Chicago Press, 314.

Armstrong, P. (2000) *The English Parson-Naturalist*. Leominster: Gracewing.

Clines, D. (2011) *Job*. Bellingham, WA: Thomas Nelson.

Davies, S., Macnaghten, P. & Kearnes, M. (eds.) (2009) *Reconfiguring Responsibility: Deepening Debate on Nanotechnology*. Durham: Durham University.

Dupuy, J.-P. (2010) "The Narratology of Lay Ethics", *Nanoethics* 4 153-170.

Geo, C. & Geo, S. (2015) *http://www.geoengineeringwatch.org*

Habermas, J. (2006) "Religion in the Public Sphere," *European Journal of Philosophy* 14, 1–25.

Heisenberg, W. (1959) *Physics and Philosophy: The Revolution in Modern Science* (transl. from earlier German edn). London: Allen & Unwin.

Latour, B. (2004) *Politics of Nature: How to Bring the Sciences into Democracy* (Transl. Catherine Porter). Cambridge, Mass: Harvard University Press.

—. (2008) "'It's development, stupid!' or: How to Modernize Modernization", in *Postenvironmentalism*,. J. Procter (ed.). Cambridge, MA: MIT Press.

McLeish, T. (2014) *Faith and Wisdom in Science.* Oxford: Oxford University Press

Newsom, C. (1996) "The Book of Job", in L.E. Keck *et al.* (eds), *The New Interpreter's Bible*, Vol. 4. Nashville: Abingdon. Quoted in Clines, *op. cit.*

Owen, R. *et al*. (2012) *Science and Public Policy* 39, 751-760.

Macnaghten, P. & Carro-Ripalda, S. (eds) (2015), *Governing Agricultural Sustainability: Global Lessons from GM Crops*. Forthcoming

Robertson, D. (1973). "The Book of Job: a Literary Study". *Soundings*, 56, 446- 468.

Robson, J.M. (1990). "The Fiat and Finger of God: The Bridgewater Treatises". In R.J. Helmstadter & B.V. Lightman (eds.), *Victorian Faith in Crisis: Essays on Continuity and Change in Nineteenth-Century Religious Belief.* Stanford, Calif.: Stanford University Press

Steiner, G. (1998) *Real Presences*. London: Faber and Faber

Chapter Three

The Theological Origins of the Concept of Laws of Nature and Its Subsequent Secularisation

John Henry

Origins: Laws as Nonspecific Causes

The concept of laws of nature seems to have had very ancient origins – Edgar Zilsel and Joseph Needham, two historians who were among the first to study the concept, claimed to find evidence of the idea among the Babylonians (Zilsel 1942, 249; Needham 1951, 301); Emile Meyerson, a renowned French philosopher of science, even talked about it in connection with "primitive man" (Meyerson 1930, 20). This seems to suggest that it is an idea that has been hard to resist. But it has always been a notion fraught with difficulties. The earliest attempts to express the idea among the Ancient Greeks, for example, show just how difficult it was to put across. The Milesian philosopher, Anaximander (*c.* 610–547 BC) suggested that "Things give satisfaction and reparation to one another for their injustice, as is appointed by the ordering of time". By "things" here he evidently did not mean things in human affairs, but things in general; he was making a comment about the nature of the world, but felt the notion of justice came close to the point he wanted to make. Similarly, Heraclitus (*floruit* 500 BC) said: "The Sun will not overstep his measures; if he does, the handmaids of Justice will find him out" (Warner 1958, 14, 26).

Obscure as these early speculations are, it seems clear that they are meant to have *explanatory* import – the aim is to explain why it is that things in the natural world behave the way they do. Anaximander seems to be suggesting that things act in a regular, ordered way, because if they did not, they would have to pay the consequences. Heraclitus is explaining why the Sun moves in an evidently restricted, and very regular, way across the heavens – because, if it did not, its erratic motions would attract the attention of the long arm of the law – not the boys in blue, but the handmaids of Justice.

The clear implication of these comments is that the Sun and Anaximander's "things" are aware in some way, as we are, of how they are supposed to behave. Plato (*c.* 427–*c.* 347 BC) pointed out that "if the stars were soulless, and consequently devoid of reason, they could never have conformed to such precise computations" (*The Laws*, 967a). The precise computations, of course, are those of the astronomers, which reveal that the motions of the stars do indeed conform to strict parameters.

It is usually said that the Ancient Greeks did not fully develop a concept of natural law, because they saw laws as established merely by convention, and it is true that the idea is not prominent (Milton 1981). There are strong foreshadowings of the idea in Plato, however, who used the perceived order and harmony of the cosmos as a model for the state. In the dialogue named after Gorgias, for example, we read:

> Wise men tell us, Callicles, that heaven and Earth and gods and men are held together… by orderliness, self-control, and justice, and that is the reason… why they call the whole of this world by the name of order [*kosmos*], not of disorder or dissoluteness (*Gorgias*, 507e).

It is hardly surprising, in the light of this, that Boethius (*c.* 480–524), who drew chiefly upon Plato's philosophy in his *Consolations of Philosophy* (*c.* 523–4), should write that the Creator's power "turns the moving sky/And makes the stars obey fixed laws" (*Legemque pati sidera cogis*) (Boethius [ca 524] 1981, 48).

But Boethius was a Christian, and he therefore also inherited the notion of laws of nature originating in the Judaeo-Christian tradition. Certainly, the Judaeo-Christian God was, among other things, a legislator who could make a decree for the rain (Job 28v26), and the sea (Proverbs 8v29), so that, for example, "the waters should not pass his commandment". Even so, laws of nature do not figure greatly in Mediaeval philosophy, presumably because of the dominance of Aristotelianism, which had no tradition of laws of nature. There are, however, scattered references throughout the period which, although vague, indicate that thinkers knew well enough what was meant by the term (Milton 1981, 1998; Oakley 1984).

It was only in the sixteenth century, when natural philosophers began to look around for alternatives to the increasingly moribund scholastic Aristotelianism, that laws of nature began to figure more emphatically in theology and natural philosophy. Perhaps the main idea here was summed up by the leading French medical writer, Jean Fernel (1597–1558):

> What God long ago started work upon by his particular actions, he has now entrusted to heaven to carry on its management, as if he were taking time off. And all that we say comes into being by the laws of nature, did first proceed from God; God certainly nowadays generates fairly few things directly without mediation by nature or seed, but regulates everything through heaven, having established nature's laws (Fernel [1548] 2005, 355).

The idea is essentially the same as, and indeed an extension of, the mediaeval view that God is the first cause, but he chooses to act through secondary causes. Instead of referring to specific natural agents however, which could act as secondary causes (such as fire, which could bring about a number of effects), Fernel refers to natural laws which are responsible for "all that we say comes into being."

This is a useful quotation for our purposes because it draws attention to two things. Firstly, that the concept of laws of nature was always intimately connected with God, the legislator and underwriter of these laws. And secondly, the laws were always seen as *causal* principles, which explained the nature of the world. Both of these are apparent in Fernel's statement, but he is simply making explicit what was always inherent in the notion. The causal nature of the laws is hinted at even in Anaximander and Heraclitus, and by the time we get to Plato, the role of the divinity is also part of the story.

But throughout all these ages, the concept of laws of nature has been used in a rather vague way, to explain "all that we say comes into being". The concept was entirely non-specific; laws seemed to simply stand in for secondary causes, whenever and wherever a specific secondary cause could not be determined (Milton 1981, 1998). Remarkably, this changed in the early seventeenth century as a result of the innovatory natural philosophy of one man – René Descartes (1597–1650). And this can be seen as one of the pivotal moments inaugurating modern science, and indeed, if we want to wax lyrical about it, modernity itself (Henry 2004).

A New Precision: Laws as Specific Causes

Descartes, of course, was the thinker who first developed a fully-worked out system of what came to be called the mechanical philosophy. This has clear affiliations to Ancient atomism – it certainly shared the atomists' assumption that all bodies were composed of invisibly small particles of matter, and that the widely different properties of bodies could be explained in terms of the combinations, arrangements, and motions of the invisibly small particles. But Descartes went further. His system was far more kinematic than Ancient atomism, with a much greater emphasis upon the motions of his particles, and of the bodies which they constituted.

This emphasis upon motion derived from the fact that Descartes wanted to develop a system of natural philosophy which had no recourse to, no need for, *occult* concepts, such as forces, or powers. Descartes believed he could explain everything, all physical phenomena, just in terms of bodies in motion. To make this work, he had to make a number of assumptions about the nature of motion, including the very important assumption that motions could be transferred from one part of the system to another – that is to say, motions had to be capable of being passed on from one particle to another, and where necessary passed back, to account for the various phenomena we see happening all around us. Descartes also recognised that, since he was explaining virtually everything in terms of the motions of particles, he could now make the laws of nature, previously so vague, precise and highly specific. His assumptions about the nature of motion and about bodies in motion, became his three laws of nature. From now on, the laws of nature were no longer simply vague terms – they were clearly stipulated and codified, and applied in a rigorous way (Henry 2004).

So, the laws of nature are the lynch-pin of Cartesian and other mechanical philosophies. All explanations in the mechanical philosophy depend upon the interactions of moving particles of matter; the laws of nature stipulate precisely how pieces of moving matter can and do interact; therefore, all explanations in the mechanical philosophy derive from the laws of nature. It is important to note, therefore, that the three laws stipulate the only ways that causal interaction takes place.

It is important to recognise that these are *causal* laws – laws 1 and 2, especially, might look, on a superficial glance, as though they are simply descriptive, and merely stating regularities in nature. Take law 2 ("That all movement is, of itself, along straight lines; and consequently, bodies which are moving in a circle always tend to move away from the centre of the circle they are describing") we might imagine that all Descartes is saying there is that bodies always move in straight lines. But in effect, what Descartes is saying is that if we want to understand how and why bodies behave as they do in all the countless phenomena of nature, we need to suppose at the outset that there is an inbuilt tendency in all bodies to move only in straight lines. So, this is a statement with explanatory import – it does not just say things *happen to* move in straight lines; it says *because* things move in straight lines, various other phenomena are inevitably going to follow. Consider the example of a sling – we can feel a stone in a sling trying to move away from the centre of rotation, and yet when we release the sling, the stone does *not* fly outwards from the centre, but follows the line of the tangent – the second law enables Descartes to explain why all this is so – why a centrifugal tendency is immediately converted to a tangential tendency on the release of the sling (Hatfield 1979).

Now, it did not take Descartes long to realize (or perhaps he believed all along) that he could not claim causal efficacy for his laws without invoking God, and so, in what historians of philosophy refer to as Descartes's "metaphysical turn" (Hatfield 1993), he began to consider the theology of Providence, and to develop a thorough concern with the theological underpinning of his natural philosophy. We can see this concern with the role of God very early on, in a letter to his friend the Minim priest, Marin Mersenne (1588–1648), written in 1628:

> … in my treatise on physics I shall discuss a number of metaphysical topics and especially the following… Please do not hesitate to assert and proclaim everywhere that it is God who has laid down these laws in nature just as a king lays down laws in his kingdom. There is no single one that we cannot grasp if our mind turns to consider it. They are all inborn in our minds just as a king would imprint his laws on the hearts of all his subjects if he had enough power to do so (Descartes [1630] 1991, 22–23).

Descartes's system was first set down in his *Le monde*, completed by 1633, and it begins with a number of chapters familiarizing its readership with the kind of approach to physics seen in Ancient atomism, It is only in chapter six that God is said to play an active, and indispensable, role in the new physics, by virtue of the fact that he establishes laws of nature from the workings of which the ordered world emerges out of chaos. In the following chapter Descartes writes "I do not want to delay any longer." He wants to tell us "what the laws of Nature that God has imposed on it [Nature] are" (Descartes [ca 1633] 1998, 24–25).

Now, one reason why Descartes required God to play such a prominent role in his physics was because of the highly unfamiliar austerity of his system of causation. To explain everything, in terms of the operation of three laws of motion, was an extremely dramatic contrast with the prevailing scholastic system of natural philosophy. Scholastic causal explanations tended to be couched in terms of the active powers of one or other of the interacting things (powers which ultimately derived from what the scholastics referred to as the "substantial form" of the thing in question). The *power* in question, of course, was all too often simply defined in accordance with the required effects – what is it about opium that causes us to go to sleep? Why, it's dormitive virtue of course (Molière [1673] 1867, Troisième Intermède, vol. 3, 483; Nadler 1998).

But it wasn't just the austerity of Descartes's laws that caused problems, but the laws themselves. Consider, for example, the initial reaction of one of Descartes's earliest converts, and the man who introduced Descartes's philosophy into England, the Cambridge Platonist, Henry More (1614–1687) (Gabbey 1982). Missing the point that God was underwriting all of Descartes's laws, More complained that, in the third law, Descartes

> … is fabricating some kind of life in that when two bodies meet, he is able to accommodate their motions so that each of them notified by the other, the one about acceleration of its motion, the other about retardation of its motion, finally agrees on the same course of motion… For Descartes himself scarcely dares to assert that the motion in one body passes into another… (quoted from Gabbey 1982, 212–13).

It is hard for us, post-Newtonians that we are, to imagine that someone would not unquestioningly accept that when billiard balls collide, a transference of motion takes place. But, as far as More was concerned, this was a very strange claim to make. After all, when the white cue ball strikes a red ball, we do not see a transference of colour – the red ball does not become pink as some of the whiteness of the cue ball is transmitted in the collision. But colour and motion are both properties of the balls – why does Descartes think motion is transferred but not colour? The answer to this question takes us back to what I said earlier about the origins of the mechanical philosophy, and Descartes's belief that he can explain everything in terms of matter in motion – but to do so he needs God to guarantee the laws.

In effect, Descartes could simply say to Henry More, that God *decided* that motions should be transferred in collisions (in strict accordance with certain rules), but he also decided that colours, and other properties should *not* be transferred in the same way. That is what law 3 is stipulating ("That a body, upon coming in contact with a stronger one, loses none of its motion; but that upon coming in contact with a weaker one, it loses as much as it transfers to that weaker body"); and of course it also conforms to our daily experience (when we see a collision of moveable bodies, it looks as though motion is transferred in a way that colours are not). Descartes does not expect More to accept the possibility of transfer of motion just on Descartes's say-so, but he assumes he will accept that God could make it look as though motion is being transferred in any collision. If Descartes were simply pointing out what happens in nature, as a mere regularity, he would not need to call it a law. He did not need a law to say that in a collision the colours of the colliding bodies are *not* transferred. No explanation of the non-transference of colour is required. But we do require an explanation as to why motions are transferred – the third law provides that explanation: the third law is a statement of how (and why) causation takes place.[1]

Descartes was immensely influential but his three laws of nature proved to be incorrect, and it was only when Newton presented his alternative laws that modern physics really took off. As a result of Newton's undeniable success, it became accepted that one of the main aims – perhaps *the* main aim – of a natural philosopher should be the discovery of the laws governing the natural world (Meyerson 1930, 19–25). But, at least in their inception, it was also accepted that the laws were guaranteed by God, and were causal laws – that is to say, the laws acted as causes, and explained why things behaved the way they did. As Descartes said in the final recension of his system, the *Principia philosophiae* of 1644, God is "the general cause of all the movements in the world" (Part II, § 36) but the laws of nature are the "secondary and particular causes of the diverse movements which we notice in individual bodies" (Part II, § 37).

The concept caught on and became characteristic of future science. And indeed it is quite likely that one reason why it was so popular among early modern natural philosophers was precisely because "laws of nature" made no sense without God, and therefore demanded acknowledgement of his existence. The devout natural philosopher Robert Boyle made the point explicit:

> It is plain that nothing but an intellectual being can be properly capable of receiving and acting by a law… And it is intelligible to me that God should… impress determinate motions upon the parts of matter, and that… he should by his ordinary and general concourse maintain those powers which he gave the parts of matter to transmit their motion thus and thus to one another. But I cannot conceive how a body devoid of understanding and sense, truly so called, can moderate and determine its own motions, especially so as to make them conformable to laws that it has no knowledge or apprehension of (Boyle [1686] 1996, 24).

---

[1] It is perhaps worth noting, *pacé* Professor Cartwright, that although Descartes explained all physical events and other physical phenomena in terms of mechanisms – so much so, indeed, that his philosophy came to be called the mechanical philosophy – he himself believed that the operation of the mechanisms had to be explained by underlying laws of nature.

The fool has said in his heart there is no God, and only a fool, Boyle is saying, could believe that inanimate bodies could "obey" laws of nature without God being there to ensure they act accordingly.

Newton himself, like Descartes and Boyle, saw the laws as a shorthand way of referring to God's way of upholding the world, or to God's "ordinary and general concourse" as Boyle put it. In the Preface to the second edition of Newton's *Principia*, for example, the editor, Roger Cotes (1682–1716), wrote that God created the world and "From this source, then, have all the laws that are called laws of nature come …" (Newton [1713] 1999, 397) Similarly, in the speculative and highly influential "Queries" that Newton added to the end of his other great book, the *Opticks*, he insisted that:

> God is able to create Particles of Matter of several Sizes and Figures, and in several Proportions to Space, and perhaps of different Densities and Forces, and thereby to vary the Laws of Nature, and make Worlds of several sorts in several Parts of the Universe (Newton [1717] 1979, 403–04).

Newton's claim that God could vary the laws of nature was a response to the claims of Spinoza (1632–1677), and Leibniz (1646–1716), both of whom (though by different routes) had arrived at what Newton saw as the dangerous conclusion that the laws of nature were absolutely necessary, and that not even God could change them (Gabbey 1996; Rutherford 1995).

As a result of Newton's success in explaining the motions of the planets, and other phenomena in the *Principia* and the *Opticks*, laws of nature became a mainstay of science, and as we've already suggested, the discovery of new laws, or the explanation of phenomena in terms of already established laws, became the main aim of the physical sciences, and became a major desideratum in medicine and the life sciences.

Laws of Nature and the Action of God

It also had its uses within theology itself. Consider, for example, the use of the concept of laws of nature as a way of solving the theological problem of evil. Robert Boyle was among the first to point to this idea:

> It may be rationally said that God – having an infinite understanding… – did, by virtue of it, clearly discern what would happen in consequence of the laws by him established in all the possible combinations of them and in all the junctures of circumstances wherein the creatures concerned in them

> may be found. And that, having… settled among his corporeal works general and standing laws of motion suited to his most wise ends, it seems very congruous to his wisdom to prefer… catholic laws and higher ends before subordinate ones, and uniformity in his conduct before making changes in it according to every sort of particular emergencies; and consequently, not to recede from the general laws he at first most wisely established to comply with the… needs of particular creatures, or to prevent some seeming irregularities (such as earthquakes, floods, famines, etc.) incommodious to them… (Boyle [1686] 1996, 161).

Although this approach to theodicy was mercilessly pilloried by Voltaire in his comic novel *Candide* (1759), it was an approach that lasted a long time. Boyle's reference to earthquakes was made long before the Lisbon earthquake of 1755 which set Voltaire off on his anti-Christian crusade, as was Alexander Pope's reference to them in his *Essay on Man* of 1734:

> Ask for what end the heavenly bodies shine,
> Earth for whose use? Pride answers, "'Tis for mine:
> For me kind Nature wakes her genial power, …"
>
> But errs not Nature from this gracious end,
> From burning suns when livid deaths descend,
> When earthquakes swallow, or when tempests sweep
> Towns to one grave, whole nations to the deep?
> "No, ('tis replied) the first Almighty Cause
> Acts not by partial, but by general laws;
> The exceptions few…" (Epistle I, Section V).

But even after the Lisbon earthquake, and *Candide*, this answer to the problem of evil, seemed powerful and evidently persuasive – perhaps further testimony to the power and influence of Newtonian science and the concept of laws of nature.

At the very end of the eighteenth century, the Anglican clergyman, Thomas Malthus (1766–1834), called upon the same ideas to oppose William Pitt's plans to reform the Elizabethan Poor Law. The fact that the population increases exponentially, while the amount of food provision, at best, can only be increased arithmetically, was an inescapable law of nature and, as such, all part of God's plan. The shortage of food, Malthus wrote, is an "all pervading law of nature… And the race of man cannot, by any efforts of reason, escape from it" (Malthus 1798, 15).

If anything, this kind of theodicy achieved even greater heights in the nineteenth century. The first systematic account of evolutionary theory to be written in English, the best-selling and highly influential *Vestiges of the Natural History of Creation* ([Chambers] 1844), justified its Malthusian, or even Draconian, picture of nature, and society, in precisely the same terms. Although the contemporarily-anonymous author of *Vestiges* did not come up with the principle of natural selection, and could only base his evolutionary theory on a supposed "law of development", which ensured the progression of life forms, he did use what he called "the unbending action of his [God's] great laws" to explain the evils that all too often afflict us.

> Everywhere we see the arrangements for the species perfect; the individual is left, as it were, to take his chance amidst the mêlée of the various laws affecting him. If he be found inferiorly endowed, or ill befalls him, there was at least no partiality against him. The system has the fairness of a lottery… ([Chambers] 1844, 377)

Although this might seem like "a dreary view of the Divine economy", as the author put it, we must suppose that "the present system is but a part of a whole, a stage in a Great Progress" ([Chambers] 1844, 384–5) – or, as Voltaire might have said, all is for the best, in the best of all possible worlds.

The anonymous author, now known to have been the popular Edinburgh publisher Robert Chambers (1802–1871), developed his evolutionary ideas as a way of solving the newly realised problem which emerged out of the comparatively new science of palaeontology – this problem was *the origin of species*. It was newly recognised as a problem because palaeontological findings were increasingly confirming that the oldest rocks contained the fossil remains only of very primitive life forms, and that it was only in successively more recent rocks, that fossil remains of more advanced creatures began to appear. The evidence suggested, therefore, that God had not created the creatures all at once, as Genesis would have it, but that creatures had appeared successively, when the time was ripe. Here's what Chambers himself writes:

> It is scarcely less evident, from the geological record, that the progress of organic life has observed some correspondence with the progress of physical conditions on the surface. We do not know for certain that the sea, at the time when it supported radiated, molluscous, and articulated families, was incapable of supporting fishes; but causes for such a limitation are far from inconceivable. The huge saurians appear to have been precisely adapted to the low muddy coasts and sea margins of the time when they flourished. Marsupials appear at the time when the surface was generally in that flat, imperfectly variegated state in which we find Australia, the region

> where they now live in the greatest abundance, and one which has no higher native mammalian type. Finally, it was not till the land and sea had come into their present relations, and the former, in its principal continents, had acquired the irregularity of surface necessary for man, that man appeared ([Chambers] 1844, 149–50).

Chambers insisted "that the organic creation was thus progressive through a long space of time rests on evidence which nothing can overturn or gainsay" (1844, 153). Accordingly, we have to reconsider "in what way was the creation of animated beings effected?"

> How can we suppose an immediate exertion of this creative power at one time to produce zoophytes, another time to add a few marine molluscs, another to bring in one or two conchifers, again to produce crustaceous fishes, again perfect fishes, and so on to the end? This would surely be to take a very mean view of the Creative Power… ([Chambers] 1844, 153)

Cosmologists, geomorphologists and geologists have already established, Chambers pointed out,

> that the construction of this globe and its associates [i.e. the other planets of the solar system] … was the result not of any immediate or personal exertion on the part of the Deity, but of natural laws which are expressions of his will. What is to hinder our supposing that the organic creation is also a result of natural laws, which are in like manner an expression of his will? ([Chambers] 1844, 154)

It is possible that Chambers took inspiration here from a letter that Sir John Herschel (1792–1871) had written to Charles Lyell in 1836, and published in 1838 as an appendix to Charles Babbage's *Ninth Bridgewater Treatise*. Referring to the origin of new species as "that mystery of mysteries", Herschel denounced Lyell's discussion in his *Principles of Geology* as being too vague, and insisted that the appearance of new species should be explained by "a natural in contradistinction to a miraculous process", because God must have operated by natural law not just with regard to Newton's physical universe (as all were agreed), but he must have acted the same way with regard to *all* natural phenomena – including the development of organic forms (in Babbage 1838, 226–27).

For Chambers, foreshadowing post-Darwinian suggestions that evolution offers a *grander* view of the Creator, this enhances the glory of God: "To a reasonable mind the divine attributes must appear, not diminished or reduced in any way, by supposing a creation by law, but infinitely exalted" ([Chambers] 1844, 156). Furthermore,

> Those who would object to the hypothesis of a creation by the intervention of law, do not perhaps consider how powerful an argument in favour of the existence of God is lost by rejecting this doctrine. When all is seen to be the result of law, the idea of an Almighty Author becomes irresistible, for the creation of a law for an endless series of phenomena … could have no other imaginable source, and tells, moreover, as powerfully for a sustaining as for an originating power. (Chambers 1844, 157–8)

Now, it should be acknowledged that the anonymous author of *Vestiges* came in for much criticism. In this context, we need only note that his Law of Development was seen by some as a substitute for the Deity. Consider this comment, for example, in a letter of 1846 written by James Thomson, engineer and brother of the more famous William (Lord Kelvin):

> I do not like the development theory as given in *The Vestiges*. It appears to me that the author substitutes for the Creator, what he calls Law; and that, if he gives his assent to God as a First Cause, he at least supposes him to be now infinitely removed from all the Works of Nature, and that everything goes on now of itself just as a clock goes after its weights have been wound up (Thomson 1912, xxxvii).

But, if this is a good point against Chambers, it might also have been levelled against Jean Fernel, who had written of God taking "time off", and, it was a criticism that was levelled at Descartes and his followers, and indeed, even at Newton. Moreover, James Thomson does concede that he agrees with the concept of law in *The Vestiges*:

> Now I am strongly impressed with the idea that a law is in itself nothing and has no power; and I can view what we call the Laws of Nature in no other light than merely as expression of the will of an Omnipresent and Ever Acting Creator. [Thomson 1912, xxxvii]

Chambers tends to be sneered at as an amateur who failed to live up to the exacting standards required of a good scientist, but when the *Vestiges* appeared on the bookshelves in 1844, there wasn't much to separate him from Darwin, and indeed it was after seeing the response to *Vestiges* that Darwin decided to turn himself from a gentleman amateur into an expert naturalist, and spent the next eight years exclusively studying barnacles. But Darwin's early notebooks show the influence of the same laws-of-nature-tradition as that upon which Chambers drew.

Consider the famous entry in the B notebook (1837-38):

> Astronomers might formerly have said that God ordered each planet to move in its particular destiny. – In same manner God orders each animal created with certain form in certain country, but how much more simple, & sublime power let attraction act according to certain law such are inevitable consequences let animals be created, then by the fixed laws of generation such will be their successors (B Notebook, 101).[2]

The sentiment is repeated at the beginning of the *Origin of Species* (1859) where Darwin provides a quotation from William Whewell's *Bridgewater Treatise* of 1833 as an epigraph to the book:

> But with regard to the material world, we can at least go so far as this—we can perceive that events are brought about not by insulated interpositions of Divine power, exerted in each particular case, but by the establishment of general laws.
>
> Whewell: *Bridgewater Treatise.*

We can see also, in the D notebook of 1838, the first expression of a sentiment that would also remain in the *Origin*:

> How far grander than idea from cramped imagination that God created (warring against those very laws he established in all organic nature) the Rhinoceros of Java & Sumatra, that since the time of the Silurian he has made a long succession of vile molluscous animals. How beneath the dignity of him, who is supposed to have said let there be light & there was light… (D Notebook, 35–36).

Again, in Notebook N (1838-39), like Chambers later, he seems to agree with Herschel's point that God should act by law in all areas, not just in physics:

> We can allow planets, suns, universe, nay whole systems of universe to be governed by laws, but the smallest insect, we wish to be created at once by special act… [All] must be a special act, or result of laws (N Notebook, 36).

In the so-called "Sketch" of 1842, Darwin wrote:

> It accords with what we know of the law impressed on matter by the Creator: that the creation and extinction of forms, like the birth and death

---

[2] Darwin's Notebooks are now most conveniently consulted at *Darwin Online*, at http://darwin-online.org.uk/EditorialIntroductions/vanWyhe_notebooks.html (accessed December 2014). This website provides transcriptions as well as photographic images of the originals.

> of individuals, should be the effect of secondary laws. It is derogatory that the Creator of countless systems of worlds should have created each of the myriads of creeping parasites and slimy worms which have swarmed each day of life on land and water on this globe (Darwin [1842] 1909, 51).

And, of course, this becomes the magnificent closing paragraph of the *Origin*:

> It is interesting to contemplate an entangled bank… and to reflect that these elaborately constructed forms, so different from each other, and dependent on each other in so complex a manner, have all been produced by laws acting around us. These laws taken in the largest sense, being Growth with Reproduction; Inheritance… Variability… a Ratio of Increase so high as to lead to a Struggle for Life, and as a consequence to Natural Selection… Thus from the war of nature, from famine and death, *the most exalted object* which we are capable of conceiving, namely, the *production of the higher animals*, directly follows. There is grandeur in this view of life, with its several powers, having been originally breathed by the Creator into a few forms or into one; and that, whilst this planet has gone cycling on according to the fixed law of gravity, from so simple a beginning endless *forms most beautiful and most wonderful have been, and are being, evolved* (Darwin 1860, 490).[3]

The same theistic view of the operation of laws of nature was reiterated by two of Darwin's earliest converts to the cause of evolution. In a letter to Darwin which, with permission, was quoted in the second and subsequent editions of the *Origin*, the novelist and advocate of "muscular Christianity", Charles Kingsley (1819–1875) wrote of God's wisdom in making "all things make themselves" (Darwin 1985, vol. 7, 379). Although Kingsley did not explicitly invoke laws, when Darwin (mis)quoted him he interpolated laws into Kingsley's comment (Darwin 1860, 481). Meanwhile, Baden Powell, in his contribution to the notorious *Essays and Reviews* of 1860, praised Darwin's contribution to natural theology by uncovering "the law of natural selection" (Powell *et al.*, 1860, 139).

It seems perfectly clear from this historical evidence that laws of nature have been seen since their adumbration before Descartes, and after their establishment as more precise statements of physical behaviour in the seventeenth century, as *causal* principles imposed upon the world by God.

---

[3] I have quoted the second edition here (1860), which explicitly mentions the Creator; the first edition simply has: "…having been originally breathed into a few forms or into one". But even this first version tacitly assumes a Creator.

When laws of nature first became crucially important to a proper understanding of how the world works, with the advent of the mechanical philosophy, and the belief that all phenomena could be reduced to the actions of bodies in motion, natural philosophers felt the need to discuss and insist upon the theological underpinnings of this concept (Oakley 1984; Henry 2004). Subsequently, such overt theological discussions seem to have become less important, but the theistic aspects of the laws remained as strong as ever—that they were not so much laws of *bodies*, but *God's laws*, and signified the fixed and regular ways that God chose to act, or chose to make bodies act, in the world, was simply taken for granted by the majority of natural philosophers, and indeed, served as further proof of God's existence and Providence.

However, we also know from the history of science that no matter how profoundly religious a scientific concept might be in its origins, it can always be adapted by atheists to serve their purposes. Even though Descartes was convinced that his system of physics would not work without God, and therefore made the ontological proof of God's existence one of his starting points, and even though he was convinced that he had to draw a strict categorical distinction between bodies on the one hand and immaterial things that were capable of thinking on the other, it did not take the atheists long to exclude God and the concept of the immaterial soul from Descartes's system, and to develop an atheistic version of the mechanical philosophy (Kors 1990). The fact is that by the end of the sixteenth century atheism was building up its own momentum, and it was not going to let theistic natural philosophy stand in its way. But, neither could it ignore the intellectual authority of the natural sciences—always seen as second only to the authority of theology itself. Natural philosophy had been seen as the handmaiden to the Queen of the Sciences, Theology, since the High Middle Ages, and so atheists could not afford to ignore it. Accordingly, any theistic natural philosophy had to be converted (or perverted?) into an atheistic version.

Given that I have argued for the theological origins of the concept of laws of nature, we might expect, therefore, to find in the historical record atheists either dismissing the concept of laws of nature, or appropriating the concept for their own ends. So, I now need to show that this was indeed the case.

The Atheist Re-imagining of Laws of Nature

We can start with a self-professed atheist who yet wanted to model his own moral philosophy on the "experimental method of reasoning", by which he certainly meant Newtonianism. This is David Hume (1711–1776), often seen as a "Newtonian philosopher", and yet one who obviously could not accept Newton's own inherently theistic view of the laws of nature. I say this notwithstanding some recent scholarly claims that Hume was not as atheistic as he first appears, and indeed that he was never a fully committed atheist (Gaskin 1993; Andre 1993). It is certainly true, as J.C.A. Gaskin has pointed out, that Hume's "abundant prudence" led him to cover "his real opinions with ambiguous irony and even, on occasions, with denials of his own apparent conclusions"; nevertheless "Hume's critique of religion and religious belief is, as a whole, subtle, profound, and damaging to religion in ways which have no philosophical antecedents" (Gaskin 1993, 313). For our purposes, therefore, Hume can count not just as a writer who seemed to his contemporaries to be an atheist, but also one who was determinedly anti-theistic (Mitchell 1986).

So, how did Hume deal with the concept of laws of nature? Significantly, he did not tackle laws of nature head-on – which would have involved him in trying to take on Newton. Indeed, in a number of places he wrote of laws of nature as an undeniable feature of the physical world. Laws of nature are exploited, for example, in Hume's rejection of miracles (Norton, 1993, 19; Gaskin 1993, 330). But if Hume did believe in the reality of laws of nature, his concept of them was very different from the Cartesian-Newtonian concept of laws as causal principles upheld by God. Hume seems to have initiated the alternative view, increasingly prominent after him, that to invoke a law of nature was simply to point to a regularity in nature—this new concept of the law of nature made no assumptions about causal connections, but merely confirmed the regular concurrence of whatever was claimed to be under the law in question. In Hume's sceptical philosophy laws of nature are no longer statements revealing underlying causes at work in the universe; instead they become nothing more than statements of constant conjunction (Rosenberg 1993, 78–79).

This diminishing of the concept of laws of nature necessarily emerges from, and is bound up with, Hume's contention that we can have no knowledge of causal connection. Insisting that we have no real evidence that there is such a phenomenon as causal connection – no real evidence that a cause can create an effect – Hume cannot subscribe to the Newtonian view of laws of nature. So, for example, Hume insists that

> even in the most familiar events, the energy of the cause is as unintelligible as in the most unusual, and that we only learn by experience the frequent *Conjunction* of objects, without ever being able to comprehend anything like *Connexion* between them. (Hume [1748] 1962, Sect VII, Pt I, § 54, 70).

According to Hume, then, it is mere habit or custom which leads us to "form the idea of power or necessary connexion", but in fact, "There appears not, throughout all nature, any one instance of connexion which is conceivable by us. All events seem entirely loose and separate" (Hume [1748] 1962, Sect VII, Pt II, § 58, 74).

This last comment is nonsense, of course. Events do not *seem* to be loose and separate; on the contrary, they seem to us to be connected. But Hume's scepticism, or we might even say Hume's determination to reject what he sees as glib theistic assumptions, will not allow him to concede that there might be a causal connection between events, because to do so might concede too much to theism. If Hume acknowledges there is such a thing as cause and effect, he might have to confront the standard Newtonian view that laws of nature are codifications of causal connections, and would then have to try to come up with a non-theistic account of how an apple, or the Moon can "obey" the laws of nature. But how can such conformity to law be understood, except by invoking a God? The only alternative would seem to be to infer, as we saw the Cambridge Platonist Henry More suggesting, that inanimate bodies are aware of what they are doing, and capable of changing their own motions. Either alternative, God or animism, was equally unacceptable to Hume, so he had to forestall being led into such discussions, and therefore insisted that there is no such thing as causal connection. Consider his account of what takes place during a game of billiards:

> The mind can never possibly find the effect in the supposed cause, by the most accurate scrutiny and examination. For the effect is totally different from the cause, and consequently can never be discovered in it. Motion in the second billiard-ball is a quite distinct event from motion in the first; nor is there anything in the one to suggest the smallest hint of the other.... When I see, for instance, a billiard-ball moving in a straight line towards another; even suppose motion in the second ball should by accident be suggested to me, as the result of their contact or impulse; may I not conceive, that a hundred different events might as well follow from the cause? (Hume [1748] 1962, Sect IV, Pt I, § 25, 29)

Let us just pause here to compare Henry More's pre-Cartesian account of collision with Hume's. More says that what we see suggests that the stationary ball that is struck by the moving ball is roused into moving itself by the contact; the image is like a sleeper being awoken and roused into activity. Hume, on the other hand suggests that what we see are just two distinct events—first one ball moving, then another. And he goes on to say, that even if the collision suggested that the first ball caused the movement of the other, there is no reason to suppose, Newton's triumphs in the *Principia* notwithstanding, that the second ball should move off "in the direction of the right line in which that force is impress'd." So here we have Hume surreptitiously denying the validity of Newton's second law. Leaving that aside, it seems to me that Henry More's account seems easier to believe than Hume's – at least More's account acknowledges our intuitive and instinctive feeling that there is some connection between the motions of the two balls – to simply deny any connection, as Hume does, surely must strike us as crazy talk. Philosophers will insist that Hume, in some strict philosophically-defined sense is correct, but it seems to me that Hume was not driven by philosophical correctness here, but by the demands of his anti-theism. And his argument, consequently, depends for its effects on rhetoric, rather than on unassailable philosophy.

The upshot of Hume's discussion is that cause can only be defined "by a customary transition", as "an object followed by another, and whose appearance always conveys the thought of that other" (Hume [1748] 1962, Sect VII, Pt II, § 60, 77). This being so, it is impossible to invoke a law of nature as a *causal* principle, explaining the behaviour of moving bodies and associated eventualities. If what we, in our unphilosophical naivety, think is cause and effect is in fact nothing more than repeatedly observed coincidence, then it follows that talk of "laws of nature" cannot meaningfully be regarded as codifying any causal relationship, but must simply be taken to be a short-hand way of referring to frequently observed regular occurrences in the natural world.

This approach now becomes typical. The atheistic alternative to regarding laws of nature as *causal* (and therefore necessarily underwritten by God), was to consider them to be nothing more than a short-hand way of referring to a perceived regularity in nature. In this atheistic tradition, physics, like Henry Ford's version of history, is just one damned thing after another.

Let us move forward nearly a hundred years to the appearance of John Stuart Mill's *System of Logic* (1843). Mill (1806–1873) was an ardent atheist and a major contributor to the philosophy of science – in which he saw himself as correcting the theistic philosophies of science of William Whewell and Sir John Herschel, both of whom, as we've already seen in passing, subscribed to the view that laws of nature implied the existence of God.

For Mill, there is no suggestion of causation in laws of nature. Consider, for example, the first definition he offers in his *System of Logic*:

> It is the custom in science, wherever regularity of any kind can be traced, to call the general proposition which expresses the nature of that regularity, a law… But the expression *law of nature* has generally been employed with a sort of tacit reference to the original sense of the word law, namely, the expression of the will of a superior (Mill 1882, 230).

So, we have two levels of generalisation from mere regularity here: firstly, it is only a *custom* of scientists to refer to some phenomena by the term "law"; secondly, the term law is used merely to designate a regularity in nature – not, we should notice to designate any causal action. Mill's acknowledgement that this term implies "the expression of the will of a superior", is inserted in a cautionary way – the term has all too often been employed in this theistic way, but of course, Mill wants to imply, we should know better than to be fooled by this. Mill goes on to insist that laws of nature are simply inductive inferences: "According to this language, every well-grounded inductive generalization is either a law of nature, or a result of laws of nature" (Mill 1882, 231). Inductive generalisations, of course, do not involve any deductive inferences about underlying causes. As Mill insists: "the expression, Laws of Nature, *means* nothing but the uniformities which exist among natural phenomena (or, in other words, the results of induction)…" (Mill 1882, 231).

Like Hume, Mill also denies the connection between cause and effect. Although he acknowledges: "The truth that every fact which has a beginning has a cause, is co-extensive with human experience" (Mill 1882, 236). He refers to this as the "Law of Causation" – implying that the belief that everything has a cause is merely based on inductive generalization. Mill fully acknowledges the importance here of deciding what is meant by cause: "The notion of Cause being the root of the whole theory of Induction, it is indispensable that this idea should, at the very outset of our inquiry, be, with the utmost practicable degree of precision, fixed and determined" (Mill 1882, 236).

It isn't long, however, before we learn that Mill's use of the term "cause" does not coincide with everyday usage:

> … when in the course of this inquiry I speak of the cause of any phenomenon, I do not mean a cause which is not itself a phenomenon; I make no research into the ultimate or ontological cause of any thing… [T]he causes with which I concern myself are not *efficient*, but *physical* causes. They are causes in that sense alone, in which one physical fact is said to be the cause of another. Of the efficient causes of phenomena, or whether any such causes exist at all, I am not called upon to give an opinion. The notion of causation is deemed, by the schools of metaphysics most in vogue at the present moment, to imply a mysterious and most

> powerful tie, such as can not, or at least does not, exist between any physical fact and that other physical fact on which it is invariably consequent, and which is popularly termed its cause: and thence is deduced the supposed necessity of ascending higher, into the essences and inherent constitution of things, to find the true cause, the cause which is not only followed by, but actually produces, the effect. No such necessity exists for the purposes of the present inquiry, nor will any such doctrine be found in the following pages (Mill 1882, 236).

What Mill means by *physical* causation, as opposed to *efficient* causation, he tells us, is simply "invariability of succession" as revealed by observation. Atheistic thinkers had long since expelled so-called "final causes" from science, but now we have Mill excluding efficient causes too – the fact is, as Mill recognised, efficient causes in nature, as much as final causes, seemed to point, at least to the majority of contemporary scientists, to a Providential God, and Mill could not go down that road. Furthermore, he tries to suggest that anyone who does go down that road is foolishly accepting as true "a mysterious and powerful tie" which "can not, or at least does not, exist".

Let me use as a final example of the atheistic view, that of T. H. Huxley (1825–1895). Again, it might be objected that Huxley was not a full-blown atheist, but merely agnostic. We need not decide upon this one way or another here; suffice it to say that Huxley only uses agnosticism as a label to apply to himself – he does not develop a distinctive "agnostic" way of arguing about Christian beliefs. Huxley's arguments, even more than Hume's, are unequivocally anti-theistic. For our purposes, therefore, he can be seen as representing atheistic trends in late Victorian Britain.

Huxley discussed the concept of laws of nature in his 1887 essay "Science and Pseudo-science", explicitly an attack on The Duke of Argyll, and on the author of the *Vestiges*, but implicitly, of course, an attack on all theistic thinking in science. The Duke in question is the eighth, George Campbell (1823–1900), and author of *The Reign of Law* (1867), a book which is entirely in the theistic tradition I have already outlined. Huxley begins by trying to suggest that the term law of nature is nothing more than a reference to regularity in nature:

> I am not aware that the term is used by… authorities in the seventeenth and eighteenth centuries, in any other sense than that of "rule" or "definite order" of the coexistence of things or succession of events in nature (Huxley 1894, 104).

Accordingly, he immediately takes the Duke to task for trying to make something more of it:

> The Duke of Argyll, however, affirms that the "law of gravitation" as put forth by Newton was something more than the statement of an observed order. He admits that Kepler's three laws [of planetary motions] "were an observed order of facts and nothing more." As to the law of gravitation, "it contains an element which Kepler's laws did not contain, even an element of causation, the recognition of which belongs to a higher category of intellectual conceptions than that which is concerned in the mere observation and record of separate and apparently unconnected facts" (Huxley 1894, 104–05).

We can sense that Huxley would have quoted the Duke's "even an element of causation" with a sneer of contempt – but as an anti-theist, he had little choice but to treat it with contempt, he could not concede causal laws without opening the door to theism. Accordingly, a little later he insists:

> Newton assuredly lent no shadow of support to the modern pseudo-scientific philosophy which confounds laws with causes. I have not taken the trouble to trace out this commonest of fallacies to its first beginning; but I was familiar with it in full bloom, more than thirty years ago, in a work which had a great vogue in its day – the "Vestiges of the Natural History of Creation"... (Huxley 1894, 108).

Huxley goes on to say that the laws in the *Vestiges* "are existences intermediate between the Creator and His works", and natural phenomena operating by law are again sneered at "as relieving the Creator from trouble about insignificant details." (Huxley 1894, 109) Huxley is now writing in 1887 and might have taken his beloved Darwin to task for making similar claims – but, needless to say he found it more than expedient to leave Darwin out of it. Meanwhile, he portrays the *Vestiges*, not as a supreme example of Victorian scientific naturalism (which is how it is usually perceived) (Secord 2001) but as a supernaturalist account of the world:

> The only hypothesis which gives a sort of mad consistency to the Vestigiarians views is the supposition that laws are a kind of angels or *demiurgoi*, who, being supplied with the Great Architect's plan, were permitted to settle the details among themselves (Huxley 1894, 110).

If we go back to Huxley's review of the *Vestiges*, originally published in 1854, we see what should now be a predictable attempt to deny any causal aspects of laws of nature, and to present them instead merely as a shorthand way of referring to particular observed regularities in nature. He begins by telling us what the author of the *Vestiges* means by law of nature:

> … we quote here the author's own words, that "The actual proposition of the 'Vestiges' is creation in the manner of law, that is, the Creator working in a natural course, or by natural means"… Here then is the idea of the book, and if the author has not demonstrated this, it is so much waste paper. There is, however, one preliminary which must be settled before passing to the question of demonstration – namely, has this potent proposition, as it is here expressed, any intelligible meaning at all? (Huxley 1854, 426).

Predictably, for Huxley the concept of natural law expressed in this way has no real meaning, even though as he admits, it is the view of natural law "which is entertained by the Vestigiarian in common with the great mass of those who, like himself, indulge in science at second-hand and dispense totally with logic" (Huxley 1854, 427). The reader should bear in mind that this also includes Whewell, Herschel, James and William Thomson (Lord Kelvin), Darwin, and many other leading scientists. According to Huxley, however, "In truth, every one who possesses the least real knowledge of the methods of science, is perfectly aware that 'natural laws' are nothing but an epitome of the observed history of the phenomena of the universe" (Huxley 1854, 429).

Conclusion: Secularism Prevails?

In concluding this brief survey, we need to acknowledge that professional philosophers of science are now treating the concept of laws of nature extremely seriously, and trying to work out just what status they can be said to have. Accordingly, there are numerous competing theories about what is meant by a "law of nature". But these are recent developments, and if we go back before the 20th Century, at least in Britain, we can see that there were effectively only two main ways of considering the status of laws of nature. Originally, the laws of nature were seen as causal principles, understanding of which explained how and why natural phenomena took place as they did. But necessarily implicit in this was a belief that God must be ensuring that inanimate bodies conform to the law-like behaviour that is required of them. There can be no denying, therefore, that those original laws were theological and theistic. What's more, from a purely scientific perspective they were extremely powerful and fruitful. We can also see that the earliest attempts to deny the theistic implications of the laws of nature were developed, unsurprisingly, by thinkers who already had a prior commitment to atheism or anti-theism. These all took a standard line in opposition to the theistic view that laws were causal principles serving an explanatory purpose, insisting instead that laws of nature were simply a shorthand way of referring to particular observed regularities in nature, had no further meaning, and explained nothing.

In spite of, or perhaps *because* of, the austerity of this secular concept of laws of nature, compared to the theistic concept, it seems now to be the dominant view of what laws of nature are. This is even the meaning which is now emphasised in the *Oxford English Dictionary*: "In the sciences of observation, a theoretical principle deduced from particular facts, applicable to a defined group or class of phenomena, and expressible by the statement that a particular phenomenon always occurs if certain conditions be present."[4] This is hardly surprising – most scientists now are secularists and consequently demand a secular concept of laws of nature. But there can be no denying that from its origins until comparatively recently it was predominantly, and highly successfully, a theological concept.

References

Andre, S. (1993) "Was Hume an Atheist", *Hume Studies* 19, 141–66.

Babbage, C. (1838) *The Ninth Bridgewater Treatise, A Fragment* (London: John Murray).

Boethius, A. ([ca 524] 1981) *The Consolation of Philosophy*, translated by V. E. Watts (Harmondsworth: Penguin).

Boyle, R. ([1686] 1996) *A Free Enquiry into the Vulgarly Received Notion of Nature* ed. E. B. Davis & M. Hunter (Cambridge: Cambridge University Press).

[Chambers, R.] (1844) *Vestiges of the Natural History of Creation* (London: John Churchill).

Darwin, C. (1909) *The Foundations of the Origin of Species: Two Essays Written in 1842 and 1844*, F. Darwin (ed) (Cambridge: Cambridge University Press).

---

[4] There is a subordinate note to this definition, which hints at an alternative view without acknowledging the implicit contradiction. The note reads: "The 'laws of nature', by those who first used the term in this sense [*sic*], were viewed as commands imposed by the Deity upon matter, and even writers who do not accept this view often speak of them as 'obeyed' by the phenomena, or as agents by which the phenomena are produced." See the *Oxford English Dictionary Online*, "law; III Scientific and philosophical uses, 17a."

—. (1860) *The Origin of Species*, 2nd edn (London: John Murray).
—. (1985–2010) *The Correspondence of Charles Darwin*, 18 vols (Cambridge: Cambridge University Press).
Descartes, R. (1644) *Principia philosophiae* (Amsterdam: Elzevier).
—. ([ca. 1636] 1998) *The World and Other Writings*, transl. & ed. S. Gaukroger (Cambridge: Cambridge University Press).
—. (1991) *The Philosophical Writings of Descartes, Vol. III: The Correspondence*, ed. J. Cottingham, R. Stoothoff, D. Murdoch & A. Kenny (Cambridge: Cambridge University Press).
Fernel, J. ([1548] 2005) *Jean Fernel's On the Hidden Causes of Things: Forms, Souls, and Occult Diseases in Renaissance Medicine*, ed. J. M. Forrester & J. Henry (Leiden: Brill).
Gabbey, A. (1982) "Philosophia Cartesiana triumphata: Henry More (1646-1671)" in T. M. Lennon (ed.), *Problems of Cartesianism* (Toronto: McGill-Queen's University Press), 171-250.
Gabbey, A. (1996) "Spinoza's Natural Science and Methodology", in D. Garrett (ed.), *The Cambridge Companion to Spinoza* (Cambridge: Cambridge University Press), 142–91.
Gaskin, J. C. A. (1993) "Hume on Religion", in D. F. Norton (ed.), *The Cambridge Companion to Hume* (Cambridge: Cambridge University Press), 313–44.
Hatfield, G. (1979) "Force (God) in Descartes' Physics," *Studies in History and Philosophy of Science*. 10, 113-40.
—. (1993) "Reason, Nature and God in Descartes," in S. Voss (ed.), *Essays on the Philosophy and Science of René Descartes* (Oxford: Oxford University Press), 259-87.
Henry, J. (2004) "Metaphysics and the Origins of Modern Science: Descartes and the Importance of Laws of Nature", *Early Science & Medicine*, 9, 73-114.
Hume, D. ([1648] 1962) *Enquiries Concerning Human Understanding and Concerning the Principles of Morals*, ed. L. A. Selby-Bigge (Oxford: Clarendon Press).
Huxley, T. H. (1854) "Vestiges of the Natural History of Creation. Tenth edition, London, 1853" *The British and Foreign Medico-Chirurgical Review* 25, 425– 39.
Huxley, T. H. (1894) *Science and Christian Tradition: Essays. Collected Essays, Volume V* (London: Macmillan).
Kors, A. C. (1990), *Atheism in France, 1650–1729. Volume I: The Orthodox Sources of Unbelief* (Princeton: Princeton University Press).
Malthus, T. (1798), *An Essay on the Principle of Population, as it Affects the Future Improvement of Society* (London: J. Johnson).

Meyerson, Émile (1930) *Identity and Reality* (London: George Allen & Unwin).

Mill, J. S. (1882) *A System of Logic, Ratiocinative and Inductive*, 8th edn (New York: Harper and Brothers).

Milton, J. R. (1981) "The Origin and Development of the Concept of the Laws of Nature," *Archives Européennes de Sociologie* 22, 173-95

—. (1998) "Laws of Nature," in D. Garber & M. Ayers (eds), *The Cambridge History of Seventeenth-Century Philosophy* (Cambridge: Cambridge University Press), 680-701

Mitchell, T. A. (1986) *David Hume's Anti-Theistic Views* (Lanham, MD: Rowman & Littlefield)

Molière, J.-B. de P. ([1673] 1867) *Oeuvres complètes de Molière*, 3 vols (Paris: Garniers Frères)

Nadler, S. (1998) "Doctrines of Explanation in Late Scholasticism and in the Mechanical Philosophy," in Garber and Ayers (eds.), *Cambridge History of Seventeenth-Century Philosophy* (Cambridge: Cambridge University Press), 513-52

Needham, J, (1951) "Human Law and the Laws of Nature," *Journal of the History of Ideas* 12, 3–30 and 194–230

Newton, I. (1979) *Opticks... Based on the Fourth Edition, London 1730* (New York: Dover)

Newton, Isaac ([1713] 1999) *The Principia*, transl I. B. Cohen & A. Whitman (Berkeley: Universty of California Press)

Norton, D. F. (1993) "An Introduction to Hume's Thought", in D. F. Norton (ed.), *The Cambridge Companion to Hume* (Cambridge: Cambridge University Press), 1–32

Oakley, F. (1984) *Omnipotence, Covenant and Order: An Excursion in the History of Ideas from Abelard to Leibniz* (Ithaca: Cornell University Press)

Powell, B., *et al.* (1860), *Essays and Reviews* (London: J. W. Parker)

Rosenberg, A. (1993), "Hume and the Philosophy of Science" , in D. F. Norton (ed.), *The Cambridge Companion to Hume* (Cambridge: Cambridge University Press), 64–89

Rutherford, D. P. (1995), *Leibniz and the Rational Order of Nature* (Cambridge: Cambridge University Press)

Secord, J. A. (2001), *Victorian Sensation: The Extraordinary Publication, Reception, and Secret Authorship of* Vestiges of the Natural History of Creation (Chicago: University of Chicago Press)

Thomson, J. (1912) *Collected Papers in Physics and Engineering* (Cambridge: Cambridge University Press)

Warner, R. (1958) *The Greek Philosophers* (New York: New American Library)

Zilsel, E. (1942), "The Genesis of the Concept of Physical Law," *Philosophical Review* 51 (1942), 245–79

CHAPTER FOUR

RELIGION IN AN AGE OF SCIENCE: THE LAWS OF NATURE AND CHRISTIAN PIETY IN THE BRIDGEWATER TREATISES

JONATHAN R. TOPHAM

First impressions matter. Nowhere is this more evident than in Charles Darwin's choice of opening words for his epoch-making *Origin of Species* (1859). Some modern editions would lead one to suppose that the work began with Darwin's autobiographical reflections concerning the *Beagle* voyage at the start of the introduction (Darwin [1859] 1996, 3). However, the work's opening words actually appeared opposite the title page (Fig. 4.1), and they were words that Darwin had selected, rather than written. Darwin's chosen epigraph concerned God's tendency to act in nature by laws, rather than by miracles:

> But with regard to the material world, we can at least go so far as this – we can perceive that events are brought about not by insulated interpositions of Divine power, exerted in each particular case, but by the establishment of general laws. (Darwin 1859, ii)

These strikingly theological words came from a work published a quarter of a century earlier, the Bridgewater Treatise on *Astronomy and General Physics, Considered with Reference to Natural Theology* (1833), written by William Whewell (1794–1866), fellow and tutor of Trinity College, Cambridge, and an increasingly prominent man of science. What prompted Darwin to choose to commence his revolutionary work in this way? What had Whewell's Bridgewater Treatise to add to Darwin's carefully crafted case for evolution?

The eight Bridgewater Treatises are now hardly household names, but they were certainly so in the middle years of the nineteenth century. Their origin, in a handsome bequest made to the Royal Society by the eccentric eighth Earl of Bridgewater, Francis Henry Egerton (1756–1829), gave them a high profile from the outset. The earl had left £8000 for the production of a work ….

> On the Power, Wisdom, and Goodness of God, as manifested in the Creation; illustrating such work by all reasonable arguments, as for instance the variety and formation of God's creatures in the animal, vegetable, and mineral kingdoms; the effect of digestion, and thereby of conversion; the construction of the hand of man, and an infinite variety of other arguments; as also by discoveries ancient and modern, in arts, sciences, and the whole extent of literature. (Whewell 1833, ix)

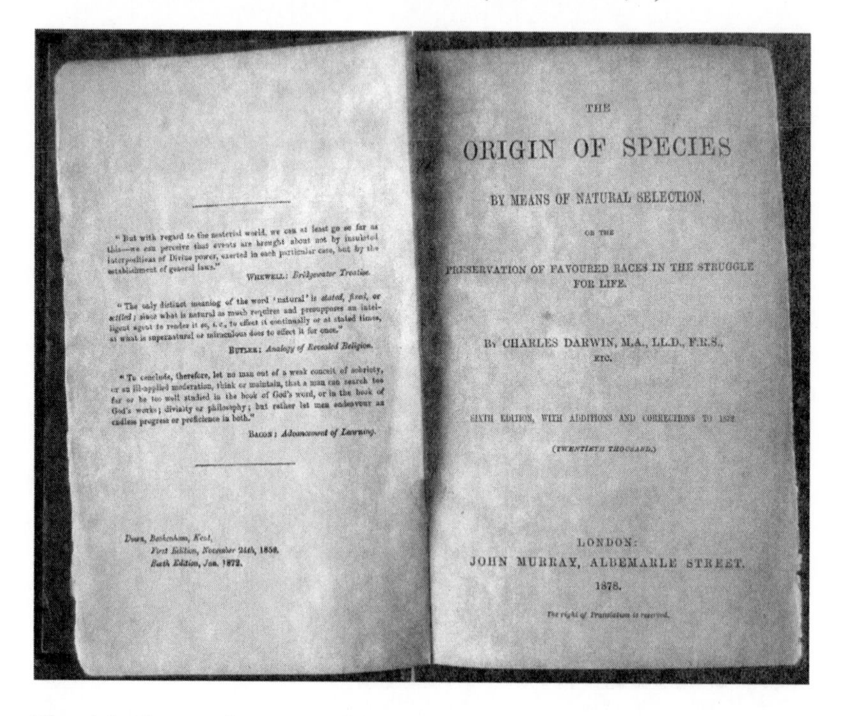

Fig. 4.1. The opening pages of the last edition of Charles Darwin's *Origin of Species* to appear in his lifetime, showing the quotation from Whewell's Bridgewater Treatise which had been there in all previous editions.

In the event, he got eight works for his money (Fig 4.2). They were written by eight prominent men of science who were appointed by the society's president with assistance from the Archbishop of Canterbury and the Bishop of London. The works produced ranged widely across the several sciences, reflecting on their religious bearings, and seeking to reassure a wide audience that developments in the sciences were consistent with Christianity. And they found their mark. Although they were far from cheap, the treatises sold well, and by 1850 over 60,000 copies of the various Bridgewater Treatises had been produced. Moreover, they were widely available in libraries, were extensively discussed in magazines, reviews, and newspapers, and were talked about from pulpits, in drawing rooms, and even in jails (Topham 1998).

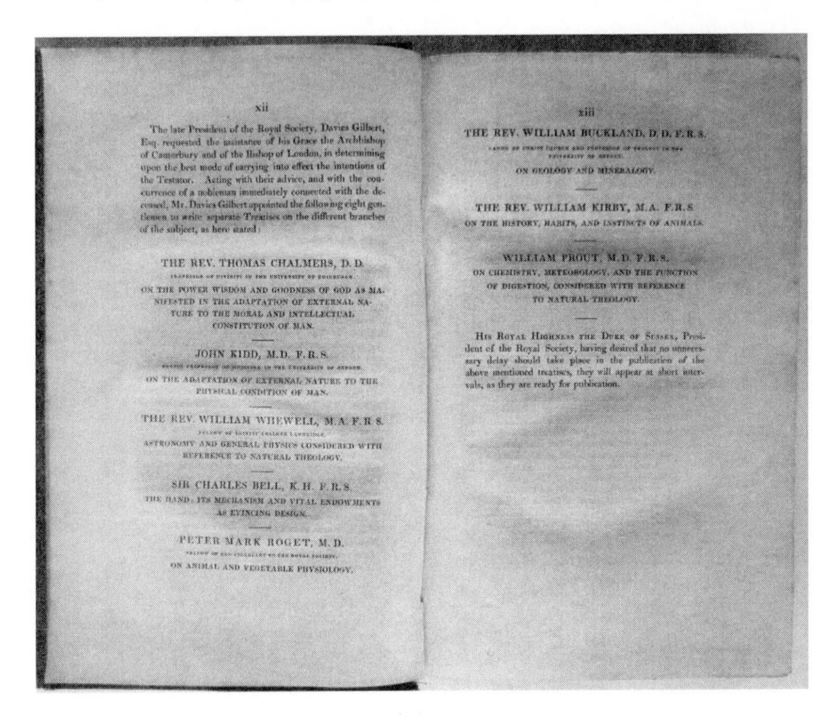

xii

The late President of the Royal Society, Davies Gilbert, Esq. requested the assistance of his Grace the Archbishop of Canterbury and of the Bishop of London, in determining upon the best mode of carrying into effect the intentions of the Testator. Acting with their advice, and with the concurrence of a nobleman immediately connected with the deceased, Mr. Davies Gilbert appointed the following eight gentlemen to write separate Treatises on the different branches of the subject, as here stated:

THE REV. THOMAS CHALMERS, D.D.

ON THE POWER WISDOM AND GOODNESS OF GOD AS MANIFESTED IN THE ADAPTATION OF EXTERNAL NATURE TO THE MORAL AND INTELLECTUAL CONSTITUTION OF MAN.

JOHN KIDD, M.D. F.R.S.

ON THE ADAPTATION OF EXTERNAL NATURE TO THE PHYSICAL CONDITION OF MAN.

THE REV. WILLIAM WHEWELL, M.A. F.R.S.

ASTRONOMY AND GENERAL PHYSICS CONSIDERED WITH REFERENCE TO NATURAL THEOLOGY.

SIR CHARLES BELL, K.H. F.R.S.

THE HAND: ITS MECHANISM AND VITAL ENDOWMENTS AS EVINCING DESIGN.

PETER MARK ROGET, M.D.

ON ANIMAL AND VEGETABLE PHYSIOLOGY.

xiii

THE REV. WILLIAM BUCKLAND, D.D. F.R.S.

ON GEOLOGY AND MINERALOGY.

THE REV. WILLIAM KIRBY, M.A. F.R.S.

ON THE HISTORY, HABITS, AND INSTINCTS OF ANIMALS.

WILLIAM PROUT, M.D. F.R.S.

ON CHEMISTRY, METEOROLOGY, AND THE FUNCTION OF DIGESTION, CONSIDERED WITH REFERENCE TO NATURAL THEOLOGY.

His Royal Highness the Duke of Sussex, President of the Royal Society, having desired that no unnecessary delay should take place in the publication of the above mentioned treatises, they will appear at short intervals, as they are ready for publication.

Fig. 4.2. Part of the announcement appended to each of the Bridgewater Treatises describing the bequest and the appointment of the eight authors.

It is perhaps easy, then, to see why the quotation from Whewell's Bridgewater Treatise appealed to Darwin as a way to commence his controversial book. Taken from one of the most successful of these widely familiar works, it seemed to give religious sanction to Darwin's attempt to provide a law-like account of the origin of new species. Indeed, Whewell's quotation appeared above another taken from the *Advancement of Learning* (1605) of Francis Bacon (1561–1626), sanctioning the view that there should be no limits to natural enquiry, but that it should proceed hand-in-hand with religious enquiry. Moreover, while Darwin later regretted having "truckled to public opinion" in using the "Pentateuchal term of creation" toward the end of *Origin*, he continued to use these epigraphs to suggest that those with theistic beliefs should have no difficulty in finding his law-like account acceptable (Burkhardt et al. 1985–, 11: 278). Indeed, in the second edition, he added a further testimony to this effect, from the *Analogy of Religion* (1736) of the bishop, theologian, and moralist, Joseph Butler (1692–1752):

> The only distinct meaning of the word "natural" is *stated, fixed,* or *settled;* since what is natural as much requires and presupposes an intelligent agent to render it so, *i.e.* to effect it continually or at stated times, as what is supernatural or miraculous does to effect it for once. (Darwin 1860, ii)

Thus Darwin's opening quotations reassured his readers that, over the preceding two centuries, the finest minds of the scientific age had been confident that law-like explanation of natural phenomena was not at odds with a continuing belief in divine action in the universe.

My interest here is with the most recent part of the history that Darwin wished to evoke. Why did he lead off with a quotation from Whewell's Bridgewater Treatise? What contribution had the Bridgewater Treatises made to discussions of natural law and divine action over the preceding quarter of a century? And why did Darwin consider that making reference to that contribution would particularly aid his cause? These are questions addressed at greater length in my forthcoming monograph about the Bridgewater Treatises, where I argue that, by seeking to demonstrate how the rapidly developing sciences might be seen to be consonant with Protestant orthodoxy, the series did much to foster a public confidence in such a consilience, which persisted through the middle third of the nineteenth century (Topham, *in preparation*). In this short chapter, I exemplify this argument by briefly examining how the treatises attempted to show that the emphasis on natural laws in modern science, far from undermining Christianity, confirmed and expanded its teachings.

The chapter begins by outlining the increasing concern expressed by many Christians in early nineteenth-century Britain, regarding the growing scope of law-like explanations offered within the several sciences. In particular, I explore the ways in which such explanations were increasingly associated with religious and political radicalism, such that leading men of science felt an increasing need to demonstrate that their approach to nature was, on both counts, safe. The second section then focuses on William Whewell's response to this situation in his

Bridgewater Treatise, both because Whewell's was the treatise that Darwin excerpted and because it was the one that developed the most sophisticated approach. I give some attention to the other Bridgewater Treatises in a third section, showing that most of the other authors were also interested to address public concerns on this topic. Finally, an epilogue reflects very briefly on the way in which the Bridgewater authors' responses to the problem of natural laws, while reassuring to many, ultimately provided license for others who, like Darwin, wished to go much further.

Natural Laws and the Religious Tendency of Science

As John Henry's chapter in this volume so nicely shows, the conception of laws of nature as manifestations of divine action had been foundational in early modern natural philosophy. By the nineteenth century, however, there was a growing sense that the expansion of explanation by natural law was potentially religiously problematic. To begin with, the standard work on natural theology in these decades – the *Natural Theology* (1802) of the Anglican archdeacon William Paley (1743–1805) – downplayed the theological importance of natural laws. Paley's central analogy between human and divine contrivance tended to be best illustrated using individual examples of mechanism. This was evident from the outset in the opening comparison of a humanly manufactured watch and a divinely manufactured eye. Thereafter his work was packed full of examples of mechanical contrivance in the organic world. Although Paley was prepared to answer objectors with the observation that a law "presupposes an agent", he had little else to say about law-like phenomena (Paley 1802, 7). Indeed, he toned down the evidential value of Newtonian astronomy, on the grounds that the analogy between the mechanism of the heavens and the mechanism of any human contrivance was, to say the least, strained (409–13). Only at the end did he briefly use laws to argue for the unity and omnipresence of the deity (478–79, 482–84).

Yet, while Paley's wonderfully accessible prose meant that around 40,000 copies of his book were in circulation by the 1830s (Fyfe 2002, 735), the same period saw dramatic changes in the sciences. The founding of a series of specialist scientific societies, beginning with the Geological Society in 1809, was just the most tangible sign of a less open, more disciplinary conception of scientific practice. Alongside this, the activities of the hundreds of mechanics' institutes that had been founded in towns and cities across the country, and of the wonderfully named Society for the

Diffusion of Useful Knowledge, suggested that scientific education was the leading priority in making the proliferating working classes of the world's first industrial nation useful and untroublesome cogs in the factory system. The founding in 1831 of the British Association for the Advancement of Science, with all the news coverage that it attracted, had been intended to confirm and consolidate the importance of scientific knowledge in public culture and national prosperity.

What did these developments mean for the accepted verities of Christianity? In his sixpenny pamphlet on the *Objects, Advantages, and Pleasures of Science* (1827), the useful knowledge society's instigator Henry Brougham (1778–1868) told his readers that the "highest of all our gratifications in the contemplations of science" was that "we are raised by them to an understanding of the infinite wisdom and goodness which the Creator has displayed in all his works" (Brougham 1827, 47). The British Association had a similar message. When its first president, the Whig MP Lord Milton, passed the torch to the president elect, Bridgewater author and geologist William Buckland, at the start of the association's second meeting – in Oxford, in June 1832 – he concluded that the "chief use" of scientific knowledge was "to lead man to lift up his mind and his heart to his Maker". Man would thus "imbibe a deeper feeling of religious awe, and acquire a stronger sense of the reverence and duty which he owes to the power, the wisdom, and the beneficence of the Creator" (Anon. 1832a, 96).

But by no means all agreed with this appraisal. The advance of science seemed to some to be the harbinger of an assault on Christianity. Speaking in Oxford, as the first of the Bridgewater Treatises went through the press, ultra-conservative Anglican clergyman and theologian Frederick Nolan (1784–1864) felt sure that the publications of the useful knowledge society, "having for their apparent object the diffusion of Science, but possessing a tendency hostile to Revelation", were clear evidence "of an organized conspiracy in active operation for the subversion of all religion". Some, he hinted darkly, might suspect that the British Association had similar intentions. In any event, while the scientific study of nature might provide evidence of the divine attributes, scientific men typically sought to account for all natural phenomena in terms of physical laws. This was tantamount to supporting atheism, and by the time one considered the damage science had done to the cause of revelation, one must be justly fearful of the consequences. A similar situation had, Nolan observed, not long before resulted in bloody revolution, a few miles across the English Channel (Nolan 1833, v–vi).

Nolan's fears may have been extreme, but they found an echo in the hopes of those at the opposite extreme. In 1821, former tin-plate worker turned radical journalist and atheist, Richard Carlile (1790–1843), had told his working-class readers that science was implacably opposed to all supernatural religion. "It is the Man of Science", he claimed, "who is alone capable of making war upon the Priest, so as to silence him effectually" (Carlile 1821, 8). Generally, the atheist radicals among Britain's working-class agitators tended to rely for their sources on such scientifically dated texts of the French Enlightenment as Baron D'Holbach's atheist *Système de la nature* (Desmond 1987). Yet, the consciousness that the likes of Carlile were poised to exploit the sciences to foment street radicalism hung over discussion of their religious implications during the tumultuous months leading up to the passage of the Reform Act in 1832.

Even those with less extreme views than Nolan had serious concerns about the consequences of the new prevalence of scientific knowledge. In the summer of 1832, the Society for Promoting Christian Knowledge reviewed the vast scope of its activity – it had circulated 1.7 million books and tracts, many of which had been used by the 740,000 pupils of Anglican National Society schools – but it was still deeply uneasy. Christianity was beset not only by the "open assaults of the infidel and the blasphemer" but by the growing tendency of the new cheap periodicals and educational works either to teach "doctrines of Materialism under the disguise of scientific principles" or "to separate knowledge from Religion, and to keep Religion altogether out of sight". In May, it reported, sales of cheap periodicals had numbered 300,000 weekly, "and of these not one was professedly engaged in the defence or support of Religion and its institutions". The secularity of scientific literature must be addressed, and the SPCK promptly launched its own cheap weekly magazine (Anon 1832b, 13–17).

It was in this context that our eight leading men of science wrote on "the Power, Wisdom, and Goodness of God as manifested in the creation". The Bridgewater authors thus had a perfect opportunity to provide their own account of the religious bearings of the latest scientific findings. In their different ways, they sought to respond reassuringly to the wide range of anxieties manifested by their contemporaries concerning the tendency of scientific knowledge. Historians have often viewed the Bridgewater Treatises as restatements and amplifications of the argument from design as found in Paley's *Natural Theology*. However, none of the Bridgewater authors saw themselves as writing a theological treatise, let alone a carefully constructed defence of the arguments of natural theology, as Paley had done. Indeed, several of the authors expressed serious reservations about the epistemological validity and/or the apologetic usefulness of natural theology, strictly so called. In this, they echoed widespread reservations about natural theology in early nineteenth-century Britain, often expressed in relation to Evangelical and High Church concerns about the primacy of the scriptures.

Rather than being interested in producing Paleyan treatises of natural theology, the Bridgewater authors viewed themselves as making contributions to a new kind of literature, which offered to a large and well educated audience an accessible account of the state of the sciences, and an interpretation of their purposes and tendency – in this case, especially their religious tendency. The emergence of a new kind of reflective work on the sciences at this period reflects a number of major developments, including the increasingly disciplinary and specialized character of the sciences, their growing public profile, and the development of an increasingly large and diverse reading public as result both of socio-cultural changes and the industrialization and commercialization of book production. As the 1820s came to an end, the British publishing market witnessed the appearance of numerous such books, including Humphry Davy's *Consolations in Travel* (1830), John Herschel's *Preliminary Discourse on the Study of Natural Philosophy* (1831), and Mary Somerville's *On the Connexion of the Physical Sciences* (1834) (Secord 2014).

It is against this backdrop, rather than against the backdrop of Paley's *Natural Theology,* that we must view both the writing and the success of the Bridgewater Treatises. For the authors themselves, these were works intended to reassure a somewhat troubled public about the religious tendency of the sciences. When they appeared, their sales confirmed that there was a market for such works. Moreover, the reactions of reviewers and readers demonstrated that their attractions lay largely in the religiously reassuring manner in which they expounded the latest findings of the several sciences.

William Whewell and the Problem with Laws

Of all the treatises, it was the first to appear, William Whewell's *Astronomy and General Physics* (1833), that was the most explicit in identifying its object as being "to lead the friends of religion to look with confidence and pleasure on the progress of the physical sciences" (Whewell 1833, vi). Whewell knew only too well that there were sincere friends of religion who were uneasy about the tendency of the sciences. The point had been brought home to him forcibly by the strongly expressed opinions of his friend from undergraduate days, the Suffolk clergyman Hugh James Rose (1795–1838). On 2 July 1826, the scarlet-clad dignitaries of the University of Cambridge had gathered in Great St Mary's Church to hear Rose deliver the sermon for Commencement Sunday. He proceeded to deliver a strongly worded critique of the irreligious tendency of the natural sciences. The modern obsession with useful, wealth-generating knowledge was putting true religion in danger, Rose claimed. Where the study of literature fostered knowledge of moral and religious truths, the study of nature had little or nothing to contribute in this pre-eminent department of knowledge, and only tended to encourage damagingly shallow habits of mind (Rose 1826).

Whewell was significantly troubled by his friend's comments. Responding later in the year to the published version of the sermon, he told Rose that he was convinced that there was nothing in the nature of experimental science "unfavourable to religious feelings". Rather, pious sentiments might be fostered by the study of science. Indeed, scientific ignorance left a person "blind to many and wonderful views which, properly considered, it gives him of the relations of ourselves and the world to the Deity" (Todhunter 1876, 2: 76, 78). Whewell had been ordained priest in May 1826, and when invited to deliver a series of afternoon sermons at Great St Mary's the following February, he told Rose that he would take the opportunity to develop this riposte. He had no interest, he insisted from the pulpit, in developing a theistic proof. Rather, he wished to manifest how modern science, properly understood, supported religious feelings, rather than undermining them as Rose suggested (Brooke 1991, 163).

Many of the arguments developed in these sermons were reworked at greater length in Whewell's Bridgewater Treatise, and the work was characterized by the same modest tone of conciliation, rather than by a strident claim for the religious value of a scientific natural theology. He had been commissioned, he asserted in the preface to his treatise, to show "how admirably every advance in our knowledge of the universe harmonizes with the belief of a most wise and good God" (Whewell 1833, vi). In particular, since "the peculiar point of view" of modern natural philosophy, was that "nature, so far as it is an object of scientific research, is a collection of facts governed by *laws*", he took as his aim the objective of showing that "this view of the universe falls in with our conception of the Divine Author" (3).

What Whewell had especially in mind was that generation of outstanding French natural philosophers, with Pierre-Simon Laplace (1749–1827) at their head, who had sought to extend and complete

Newton's law-like account of the universe, but who had done so without reference to divine action. During Whewell's early career, Laplace's growing reputation as an atheist, determined to exclude God from nature, had played increasingly to the fears of science's critics. Several months before Rose's commencement sermon, Whewell's Sabbath reading – a passage from the popular devotional aid, *Reflections on the Works of God* (*Betrachtungen über die Werke Gotte*, 1772–76) by German pastor Christoph Christian Sturm – had set him reflecting on the topic. Sturm had observed that the migration of birds testified that the universe had been arranged with great wisdom. Whewell now questioned why, to some, the advance of law-like explanation seemed to undermine a belief in providence. Laws were properly understood as the "rules of operation" of the creator, and knowledge of them provided the Christian with "a nearer and nearer knowledge of the thoughts of God as they regard the material world" (Todhunter 1876, 1: 360–65).

Whewell developed this view in replying to Rose in his 1827 sermons, and he did so at greater length in his Bridgewater Treatise. Introducing his readers to natural laws, he promised to shed light on the character of the creator from the character of his administration. This was evident, he insisted, both in the "constitution" of laws and in their "combination" to make the course of nature what it was (Whewell 1833, 5). Since the sciences that had been most successfully reduced to laws were astronomy and meteorology, the analysis was to revolve around these two. Without further deliberation, Whewell proceeded to fill two-thirds of his treatise with a survey of adaptations in the laws of nature. The first of his three "books" concerned the "Terrestrial Adaptations" found in the relation between astronomical and meteorological laws and the properties of plants and animals; the second concerned the "Cosmical Arrangements" by which heavenly objects had been created and preserved. Many of the examples in the first were relatively conventional instances of adaptation, but it was in the second that the capacious reassurance of Whewell's emphasis on laws came into play.

In describing the laws by which the universe operated, Whewell's concern was to "trace indications of the Divine care" (150). Thinly veiled behind this stood the spectre of Laplace, whose sneering scepticism became Whewell's foil. Laplace had shown that the solar system was stable, and consequently did not (as claimed by Newton) require God's intervening adjustments. Whewell asked: "Now is it probable that the occurrence of these conditions of stability in the disposition of the solar system is the work of chance? Such a supposition", he continued, "appears to be quite inadmissible" (165). The point became most explicit when

Whewell examined Laplace's own explanation for the "*primitive cause*" of this stability. In a chapter headed "The Nebular Hypothesis", he described Laplace's attempt to account for the origin of the solar system naturalistically in terms of the condensation of nebulous matter under the influence of gravity. Whewell made it clear that he considered the physical case for the nebular hypothesis to be unproven. However, his concern was to demonstrate that such a law-like explanation, if established, would not undermine the Christian's belief in a creator. Rather, it would transfer "our view of the skill exercised, and the means employed to another part of the work" – to the laws and initial conditions that allowed such a development to occur (184). The Christian had nothing to fear, this extreme example made clear, from the advance of science.

Having thus shown the harmony of modern science with Christianity, Whewell devoted the final book of his treatise ("Religious Views") to a somewhat miscellaneous series of important and wide-ranging reflections. Many of these themes had appeared in his sermons, and the short chapters were like a series of homilies. Especially important from the perspective of this volume are Whewell's further reflections on the religious tendency of natural laws. Where the first two books of the treatise had examined the adaptedness of nature's laws, he now reflected that, to most people, the very existence of natural laws in itself implied "a presiding intelligence" (296). How did we come to this impression? Whewell was reluctant to commit himself. The "various trains of thought and reasoning" which led people from the natural world to God typically did not involve "any long or laboured deduction", he observed. The impression was "so widely diffused and deeply infixed" that some had wondered whether it was perhaps "universal and innate" (293–94). Leaving the question unanswered, Whewell was content to rely on history. The impression that "law implies mind" had given rise to the "natural religious belief" of our species, he claimed (300). Moreover, the impression had continued to operate in the scientific age. Those who had been involved in scientific discovery, he argued at length, had been "peculiarly in the habit of considering the world as the work of God" (308).

The reassuring intent of this analysis became clear in Whewell's next chapter, which was the one on which he most wanted to hear his friends' comments. With his eye firmly on Rose, he observed,

> Complaints have been made, and especially of late years, that the growth of piety has not always been commensurate with the growth of knowledge, in the minds of those who make nature their study. … The opinion that this is the case, appears to be extensively diffused, and this persuasion has

> probably often produced inquietude and grief in the breasts of pious and benevolent men (323).

Whewell considered such claims exaggerated, but sought in any case to explain how the likes of Laplace could have fallen into irreligious views. The danger lay in the "deductive habits" of many eminent scientific men, who, since they were engaged in unfolding the consequences of previously discovered laws, came to view those laws as necessary. By contrast, the "inductive habits" of scientific discoverers made them conscious that they had deciphered a law that might have been otherwise, but for the will of the legislator.

In a chapter on "Final Causes" Whewell asserted that design was something perceived immediately, rather than being the product of a train of reasoning. Using the language of Immanuel Kant (1724–1804), he described it as a "regulative principle" (345–46), by which humans make sense of the phenomena they experience. Only those, like Laplace, who had indulged in a peculiar mental discipline could throw off the impression of final causes, or suggest that it amounted to an ignorance of real causes that would be banished by the extension of natural law. For those whose minds were properly constituted, scientific research merely transferred design "from the region of facts to that of laws" (349).

Whewell thus provided a thorough-going, sophisticated, and, above all, pastorally sensitive account of how the natural laws that were ever more the substance of the sciences related to Christian piety. The laws that science uncovered were entirely consistent with the administration of a wise and good creator. The expansion of law-like explanation, even in the history of the creation, only moved the believer's sense of divine agency from one place to another. The very existence of laws in nature implied the existence of a divine mind, and anyone engaged in inductive science would always be alive to the contingency of natural laws, and their dependence on the divine will. Only those exclusively devoted to deductive science were in danger of losing sight of God in necessitarian atheism, Whewell concluded. And many of his readers were impressed, considering that he had done much to demonstrate that the scientific and the religious temper were mutually supportive. Little wonder, then, that Darwin later sought the support of Whewell's Treatise when he wrote the *Origin of Species*.

Laws in Chemistry, Morphology, and Geology

Darwin's use of Whewell's distinctive analysis of the religious significance of natural laws gives it particular historical piquancy. However, Whewell was not the only Bridgewater author for whom the expansion of law-like explanation was a matter of concern. Nor, indeed, was he the only one whose statements on the subject were later appropriated by more radical thinkers for reassuring purposes. None of the other Bridgewater authors brought Whewell's philosophical sophistication to the subject, but their discussions are nonetheless revealing in illustrating how central the question of laws was in public discussion about the religious tendency of the sciences, and how leading scientific men sought to address it.

Arguably the most philosophically sophisticated of the remaining authors was the Church of Scotland minister, Thomas Chalmers (1780–1847), whose treatise came out just weeks after Whewell's. Chalmers had been appointed Edinburgh's Professor of Divinity in 1827, but as a young man he had nurtured serious scientific ambitions, and his *Astronomical Discourses* (1817), delivered from his Glasgow pulpit to a fashionable audience, had created a national sensation and given him a wide reputation as a writer on the religious bearing of the sciences. He was, moreover, an important writer on the science of political economy, and his treatise on "the adaptation of external nature to the moral and intellectual constitution of man" dealt largely with social and economic matters. However, his introductory chapter reviewed aspects of the logic of the design argument, and devoted an extended discussion to the place of natural laws within it, which was later republished in his more extended *Natural Theology* (1836).

Drawing on the work of the natural philosopher John Robison (1739–1805), who had taught him as a student in Edinburgh thirty years previously, Chalmers sought to emphasize the distinction between "Laws of Matter" and what he called the "Dispositions of Matter" (Chalmers 1833, 1: 16). In terms reminiscent of Whewell, he pointed out that natural phenomena could never be reduced merely to laws. The mechanism of the solar system depended on the many dispositions of matter "fixed at the original setting up of the machine" (17). Yet, Chalmers continued, atheists such as Laplace tended to focus only on laws, seeking to make them wholly responsible for the result. According to Laplace, the movement of the planets "first in one direction, second nearly in one plane, and then in nearly circular orbits" were consequences of the laws of matter, rather than separately caused dispositions of matter (18). So far, the analysis paralleled Whewell's, but Chalmers was less sanguine than Whewell about the consequences of such theorizing. If Laplace's theory were true, it would, he felt, weaken the argument from design in astronomy, although there would still be *some* argument to be made from the initial dispositions.

Not that Chalmers was downhearted. While the expansion of law-like explanation might diminish the evidence for design in astronomy, there was no shortage of evidence in the bodies of living creatures, where it was "quite palpable" that it was "in the dispositions of matter more than in the laws of matter, where the main strength of the argument" lay (19). And it was precisely in these dispositions, rather than in laws, that the analogy between a divine and a human designer operated. If the current state of the planet were destroyed, the laws of nature which sustain it could never reproduce it, independently of "the fiat and finger of a God" (27). This view, he considered, was nicely confirmed by the findings of geology. Thus, while Chalmers considered that an argument for divine existence could be mounted on the basis of natural laws, the argument based on the dispositions of matter was a "million-fold" more intense (30). And when "philosophical discovery" seemed to "enfeeble the argument for a God" by reducing two laws to one, the effect was illusory, since the result was to increase the number of accompanying dispositions necessary to account for the phenomena (48).

While Chalmers found natural laws less edifying than Whewell, he thus shared the latter's belief that their advance need not be a source of religious disquiet. However, the chemist and physician, William Prout (1785–1850), was more ambivalent. In his strangely packaged Bridgewater Treatise on *Chemistry, Meteorology, and the Function of Digestion*, Prout declined to expound the argument from design at any length, observing that divine design seemed evident to all those whose minds were not "obtuse" or "strangely constituted". But he paused to note the existence on the one hand of Humean sceptics and on the other of ...

> those who, denying a First Cause, affect to believe, that all the beautiful adaptations and arrangements, we see around us, are the result of what they call, "the necessary and eternal laws of nature,;" and who, in fact, are Atheists, or rather Pantheists, "to whom, the laws of nature are as gods".

The latter, he suggested, would be adequately answered by the "facts" in his book (Prout 1834, 173–75).

As his account proceeded, however, it became clear that Prout felt torn about natural laws. Beginning his section on chemistry with a philosophical discussion based on John Herschel's recent *Preliminary Discourse on the Study of Natural Philosophy* (1830), he laid out a radical ambition to reduce chemistry to quantitative laws. The advent of Daltonian chemistry had provided much-increased scope for such an approach, and

Prout used his treatise to elaborate his own radical hypotheses concerning atomic and molecular structure, which seemed to increase the prospect of obtaining quantitative laws. Cutting against this, however, was Prout's sense that law-like explanations in science undermined religious sensibilities. When phenomena could be readily explained in terms of causal, mechanical laws, Prout argued, then the Deity ...

> appears almost too obviously to limit his powers within the trammels of necessity; but when he operates through the medium of chemistry, the laws of which are less obvious, and indeed for the most part unknown to us, his operations have much more the character of those of a free agent, and, in many instances also, appear of a higher order, and are more striking and wonderful.

While "obvious mechanism" provided clear evidence of design, it could not arrest the attention nearly so well as the use of "means utterly above *our* comprehension", nor could it generate such "exalted notions" of divine wisdom and power (11–12).

Prout thus vacillated between championing the extension of law-like explanation in chemistry and delighting in instances that baffled the scientific imagination, such as the operation of vital forces in organic chemistry, or seemed to be instances of general laws being contravened, such as the "anomalous" expansion of cooling water as it approaches its freezing point (251). On the one hand, the regularities of nature seemed in themselves to be indications of a divine designer; on the other, laws were "infringed", just where their infringement was necessary (360). For all Prout's scientific instincts to the contrary, the latter seemed simply too rhetorically useful to neglect.

The same sense that the expansion of law-like explanation was a defining characteristic of modern science, and that its implications for religious belief needed further consideration, also motivated the authors of the treatises on the sciences of living beings, although generally to a lesser extent. One important aspect of their concern falls oddly on modern ears. A distinctive feature of the life sciences in early nineteenth-century Europe was the attempt – born out of the Romantic movement of late eighteenth-century Germany – to develop an approach that would explain animal morphology in terms of fundamental laws of organization. On the Continent, this project was often associated with both materialism and species transmutation, most notably in the work of French naturalist Etienne Geoffroy Saint-Hilaire (1772–1844). Moreover, the appropriation of such "philosophical anatomy" in London during the 1820s has been identified with radical reformism in the city's medical schools (Desmond

1989). In the hands of some of the Bridgewater authors, however, the new morphological laws were given an idealist twist, and were thus made to be consistent with Christianity.

This was especially evident in the large treatise on *Animal and Vegetable Physiology* written by physician and Royal Society secretary Peter Mark Roget (1779–1869), following his attendance at the anatomy lectures of one of the most radical of London's medical teachers, Robert Edmond Grant (1793–1874). Roget's treatise is a fascinating blend of two approaches to design in nature. In his first chapter, headed "Final Causes", Roget explored the scientific character of physiology. While some of the laws of physiology related to "the simple relation of cause and effect", he claimed, others were "founded on the relation of means to an end" (Roget 1834, 1: 22). These latter laws presupposed "intention or design", and thus Roget could follow Paley's logic in finding physiology an excellent source of functionalist arguments for God's existence. In his second chapter, however, Roget supplemented this account with one drawing on the new philosophical anatomy. Animal forms reflected a pair of countervailing laws: a law of "conformity to a definite type" and a "law of variety". By this, Roget meant that individual species in the same class were constructed on common morphological principles, but that these were applied in a wide variety of ways. Thus the same basic osteological elements were to be found endlessly varied in the whale, the bat, and the human. Such resemblances could not, Roget insisted, be explained in purely functional terms. They needed to be understood in relation to the new laws that philosophical anatomists had discovered.

By endorsing the morphological laws of Geoffroy Saint-Hilaire, Roget considered that he had provided his readers with a new source of evidence of divine action in the universe. Such laws, his treatise insisted, reflected the comprehensive plan of the creator in the living world, in a way that had not previously been apparent. The taxonomic classes represented "parts of one general plan" which "emanated from the same Creator" (52). Of course, Roget was sure to separate Geoffroy's laws from his transmutationist views, which were given short shrift. The laws of morphology were securely located in the ideal, rather than material realm. Indeed, for others of the Bridgewater authors, concern about Geoffroy's transmutationism was such that they were much less prepared to accommodate his morphological laws. Nevertheless, surgeon Charles Bell (1774–1842), physician John Kidd (1775–1851), and clergyman-naturalist William Kirby (1759–1850), all made more muted capital out of the new evidence that such laws offered for the unity of design in their Bridgewater Treatises (Bell 1833, Kidd 1833, Kirby 1835).

Perhaps surprisingly, in view of his strictly entomological pursuits, it was the septuagenarian Kirby who was arguably the most outspoken of the Bridgewater authors on the implications of natural laws for the religious tendency of the sciences. True to his principles as a High Churchman, Kirby had no time for a rationally constituted natural theology. Rather, his bulky treatise provided a review of the natural history of animals that conveyed a sense of God's well-designed plan, interspersed with occasional homilies drawn from animal life. His introduction, however, was a different matter. Drawing on some of his parochial sermons, Kirby addressed himself to what he considered to be the diseased condition of modern science: by substituting natural laws for God, and by ignoring the Bible, some men of science had, he claimed, gone seriously astray.

Like others of the authors, Kirby had especially in his sights the troubling duo of Laplace and transmutation theorist Jean-Baptiste Lamarck (1744–1829). The pair were the more disquieting on account of their obvious scientific prowess and also of their dangerous Frenchness, with all its revolutionary and atheist connotations in the decades after the French Revolution. Kirby first explained that Laplace had substituted natural causation for divine action in accounting for the origin of the solar system in his nebular hypothesis. Laplace's "Author of Nature" was "perpetually receding, according as the boundaries of our knowledge are extended; thus expelling, as it were, the Deity from all care or concern about his own world" (Kirby 1835, 1: xxii). But it was the esteemed naturalist Lamarck who was the subject of Kirby's special ire.

Kirby had first planned to answer Lamarck's transmutation theory almost a quarter of a century earlier. In the end, Charles Lyell (1797–1875) had forestalled him in the second volume of his *Principles of Geology* (1832), and Kirby was happy to direct his readers to that detailed critique. But most fundamental for Kirby was the way in which the Frenchman seemed to set up nature as God's "viceregent". In Lamarck's work, natural laws and the "*Order of Things*", effectively stripped God of his powers (xxxiv, xxxvi). This was deliberate, he argued:

> The great object both of La Place and Lamarck seems to be to ascribe all the works of creation to *second* causes; and to account for the production of all the visible universe, and the furniture of our own globe, without the intervention of a *first* (xxiv).

Many would have agreed with Kirby's diagnosis, but few with his prescription. Kirby concluded his introduction with an elaboration of the mystical system of biblical interpretation that he considered was the proper basis for conducting scientific enquiry. Only such a procedure, revealing the hidden symbolism of the cherubim in Solomon's temple, could lead to a properly ordered understanding of the powers that ruled in nature under God. It was, he quoted from the French of Heinrich Moritz Gaede, "Bible in hand that we must enter the august temple of nature, in order to understand the voice of the creator properly" (ii; translation mine).

Such concerted Biblicism was hardly a typical recourse for men of science seeking to reassure an anxious public about the religious tendency of the sciences in the 1830s. The Oxford Professor of Geology, William Buckland (1784–1856), moved swiftly in his Bridgewater Treatise on *Geology and Mineralogy* to assert both the independence of scientific enquiry from biblical interpretation and the ultimate congruence of scientific and biblical truths, dispensing with any further need to consider such questions. Instead, the bulk of his work was taken up with synthesizing an accessible overview of the history of the earth, based on a generation of work in stratigraphy, geological dynamics, and palaeontology.

The vision of earth history Buckland evoked was one of a progressively cooling globe, becoming gradually readied for human habitation through natural processes that involved a series of catastrophic upheavals correlating with the breaks between geological formations. Speculating that the globe might have begun in a fluid or nebular condition, Buckland linked this to Laplace's nebular hypothesis, which he considered the "most probable theory" of the origin of the solar system. Such law-like explanation, Whewell had shown, "would tend to exalt our conviction of the prior existence of some presiding Intelligence" (Buckland 1836, 1: 40). And, while Buckland had none of Whewell's philosophical sophistication, he was happy to apply such a perspective to the laws that accounted for the earth's physical history more generally. In his conclusion, he observed of the material elements of which the earth is composed:

> If the properties imparted to these Elements at the moment of their Creation, adapted them beforehand to the infinity of complicated and useful purposes, which they have already answered, and may have further still to answer, under many successive Dispensations in the material World, such an aboriginal constitution so far from superseding an intelligent Agent, would only exalt our conceptions of the consummate skill and power, that could comprehend such an infinity of future uses under future systems, in the original groundwork of his Creation (580).

Thus, Buckland reassured his readers, laws could never undermine our admiration of God's action in the universe.

Buckland's law-like account of the physical history of the planet suited his theological purposes well. Not only did it provide wonderful evidence of God's prospective design, in preparing the planet for human habitation, but it yielded compelling evidence that the earth's progressive history had a determinate beginning; it was not, as some atheists claimed, eternal. But Buckland had another card to play. While geological catastrophes might be natural events, the new species that populated the globe at the start of each new geological epoch were due to "the direct agency of Creative Interference" (586). The physical globe had only one miraculous beginning, but life on earth had many, and geology had broken new ground in providing compelling evidence for God's agency in the universe. Thus, while Buckland was happy to extend law-like explanation to the history of the earth, one of the primary objectives of his treatise was to amass evidence against the law-like appearance of new species as the result of transmutation.

Epilogue: On Route to the *Origin*

The use that Darwin made of William Whewell's Bridgewater Treatise in the *Origin of Species* prepared us at the outset to find in this series some serious attempts to reassure readers concerning the religious tendency of the law-like nature of modern science, including the increasingly law-like account of the history of creation. What we have seen is that, with greater or lesser degrees of sophistication and success, the Bridgewater Treatises generally attempted to do just that.

Whewell was the most explicit of the authors concerning his desire to protect religious feelings from any scientific danger. His object was to provide a means of maintaining a lively sense of divine action in a law-like universe and to reassure readers that there was nothing amiss with the idea of God creating by means of natural laws. But other authors, too, were keen to help their readers retain that sense. Chalmers echoed Whewell's views, albeit without the latter's emphasis on the argument that "law implies mind". Prout's feelings were more mixed: he argued that new chemical laws would further impress Christians with a sense of divine design, but insisted that there would always be exceptions to make God's action seem more impressive. Roget and others were keen to suggest that the morphological laws that some Continental anatomists wished to use to bolster transmutation actually provided a new kind of manifestation of the comprehensiveness of the divine plan. Kirby was also conscious of new morphological evidence for God's comprehensive plan, but was very alive to the danger that the extension of natural law posed to Christianity in the wrong hands. Only a biblically informed approach to nature could ensure that laws were kept in their proper place. Finally, Buckland suggested that the notion that the earth and its rocks had emerged through the operation of natural laws only served to emphasize the impressiveness of God's plan, but was confident that the appearance of new species fell outside such laws.

In these various ways, the Bridgewater Treatises offered readers of the 1830s and afterwards reassurance that the scientific project to extend law-like explanation in the universe was not at odds with Christianity. And while they were by no means universally or equally popular, there is clear evidence that they contributed to such a consilience. However, as Darwin's use of Whewell's work at the start of *Origin of Species* points up, the story has a twist in its tail. The Bridgewater authors were implacable in their opposition to the transmutation of species. Yet, from as early as 1837, readers of the Bridgewater Treatises applied the logic of the position outlined by Whewell and others to argue that the law-like origin of new species was also consistent with divine creation.

The first to do so was mathematician Charles Babbage (1791–1871) in his fragmentary *Ninth Bridgewater Treatise* (1837). Offended by Whewell's comments about the effects of "deductive" habits of mind, Babbage undertook to write an unofficial addition to the series. There, he suggested that new species might only be *apparent* exceptions to natural laws, which had been foreseen and foreordained by the creator:

> To call into existence all the variety of vegetable forms, as they become fitted to exist, by the successive adaptations of their parent earth, is undoubtedly a high exertion of creative power ... But ... to have foreseen all these changes, and to have provided, by one comprehensive law, for all that should ever occur, either to the races themselves, to the individuals of which they are composed or to the globe which they inhabit, manifests a degree of power and of knowledge of a far higher order (Babbage 1837, 44–46).

Babbage had a striking new means of illustrating how such extensions to natural laws might work. He pointed out that he could set his famous calculating engine to count according to one number series for an inordinate number of steps before suddenly switching to another series. It was as though the law of its operation had changed. By analogy, the God who had programmed the laws of the universe might be expected to have foreordained truly surprising changes in the natural order.

Others soon went further. In 1844, the anonymous best-seller, *Vestiges of the Natural History of Creation,* laid out an account of the evolution of the universe, from nebulae to humans and beyond, that was couched in the same language of creation by natural law that Whewell and others of the

Bridgewater authors had developed a decade earlier. The work even quoted from Buckland's treatise, as well as from Babbage's, in defending its approach. Penned by the Edinburgh publisher, Robert Chambers, *Vestiges*' enormous notoriety contributed to give the issue of evolution a central position in British public culture in the middle decades of the nineteenth century (Secord 2000, 2). It also made abundantly clear to the Bridgewater authors and other scientific men, as Pietro Corsi so ably expresses it, that "the machinery of natural laws which they had contributed to build was escaping the control of its Christian masters" (Corsi 1988, 269). In fact, Whewell had already begun expending copious amounts of ink in shoring up his position against such a possibility. In his ground-breaking writings on the history and philosophy of the inductive sciences, he argued that sciences involving historical causes, such as geology or nebular astronomy, were profoundly limited in what they could hope to achieve. In particular, he claimed, they could never yield natural accounts of origins. Whewell's response to *Vestiges* was to reissue some of these writings, together with extracts from his Bridgewater Treatise (Hodge 1991).

In view of such developments, Darwin's appropriation of Whewell's Bridgewater Treatise can hardly have come as a great surprise to its author. On receipt of a presentation copy, Whewell replied: "probably you will not be surprized to be told that I cannot, yet at least, become a convert to your doctrines" (Burkhardt 1985–, 8: 6). Four years later, and not long before his death, Whewell explained why he remained unpersuaded in a preface to the seventh edition of his Bridgewater Treatise. Darwin's theory was, he argued, founded on unsubstantiated assumptions that made it speculative. However, if it proved to be true, then the same argument about natural laws that had rendered the nebular hypothesis safe would again apply: divine design would be no less apparent (Whewell 1864, xvi–xviii). To this extent, Darwin's epigraph must have been justified in Whewell's eyes. In any case, the account of the religious bearing of natural laws found both in Whewell's and in the other Bridgewater Treatises, to some extent smoothed the path for the very transmutationist thinking that the authors had been as one in resisting.

References

Anon. (1832a). "Proceedings of the General Meeting." In *Report of the First and Second Meetings of the British Association for the Advancement of Science; at York in 1831, and at Oxford in 1832: Including Its Proceedings, Recommendations, and Transactions*. 95–110. London: John Murray.

Anon. (1832b). *Report of the Society for Promoting Christian Knowledge.* London: J. G. & F. Rivington.

Babbage, C. (1837). *The Ninth Bridgewater Treatise: A Fragment.* London: John Murray.

Bell, C. (1833). *The Hand: Its Mechanism and Vital Endowments as Evincing Design*. London: William Pickering.

Brooke, J. H. (1991). "Indications of a Creator: Whewell as Apologist and Priest." In *William Whewell: A Composite Portrait*, ed. M. Fisch & S. Schaffer. 149–73. Oxford: Clarendon Press.

[Brougham, H. P.]. (1827). *A Discourse of the Objects, Advantages, and Pleasures of Science*. London: Baldwin, Cradock, & Joy.

Buckland, W.; (1836). *Geology and Mineralogy, Considered with Reference to Natural Theology*. London: William Pickering.

Burkhardt, F., *et al.* (1985–). *The Correspondence of Charles Darwin*. 21 vols. Cambridge: Cambridge University Press.

Carlile, R. (1821). *An Address to Men of Science; Calling upon Them to Stand Forward and Vindicate the Truth from the Foul Grasp and Persecution of Superstition; and Obtain for the Island of Great Britain the Noble Appellation of the Focus of Truth; Whence Mankind Shall be Illuminated, and the Black and Pestiferous Clouds of Persecution and Superstition be Banished from the Face of the Earth; as the Only Sure Prelude to Universal Peace and Harmony among the Human Race, in Which a Sketch of a Proper System for the Education of Youth is Submitted to Their Judgement*. London: Richard Carlile.

Chalmers, T. (1833). *On the Power, Wisdom, and Goodness of God as Manifested in the Adaptation of External Nature to the Moral and Intellectual Constitution of Man*. 2 vols. London: William Pickering.

Corsi, P. (1988). *Science and Religion: Baden Powell and the Anglican Debate, 1800–1860*. Cambridge: Cambridge University Press.

Darwin, C. (1859). *On the Origin of Species by Means of Natural Selection; or, The Preservation of Favoured Races in the Struggle for Life*. London: John Murray.

—. ([1859] 1996). *On the Origin of Species*. Ed. G. Beer. Oxford & New York: Oxford University Press.

—. (1860). *On the Origin of Species by Means of Natural Selection; or, The Preservation of Favoured Races in the Struggle for Life*. 2nd edn. London: John Murray.

Desmond, A. (1987). "Artisan Resistance and Evolution in Britain, 1819–1848." *Osiris* 2nd ser. 3: 77-110.

—. (1989). *The Politics of Evolution: Morphology, Medicine, and Reform in Radical London*. Chicago & London: Chicago University Press.

Fyfe, A. (2002). "Publishing and the Classics: Paley's *Natural Theology* and the Nineteenth-Century Scientific Canon," *Studies in History and Philosophy of Science* 33: 733–55.

Hodge, M. J. S. (1991). "The History of the Earth, Life, and Man: Whewell and Palaetiological Science." In *William Whewell: A Composite Portrait*, ed. M. Fisch & S. Schaffer, 255–88. Oxford: Clarendon Press.

Kidd, J. (1833). *On the Adaptation of External Nature to the Physical Condition of Man, Principally with Reference to the Supply of His Wants, and the Exercise of his Intellectual Faculties*. London: William Pickering.

Kirby, W. (1835). *On the Power, Wisdom, and Goodness of God as Manifested in the Creation of Animals and in Their History, Habits, and Instincts*. 2 vols. London: William Pickering.

Nolan, F. (1833). *The Analogy of Revelation and Science Established in a Series of Lectures Delivered before the University of Oxford, in the Year MDCCCXXXIII, on the Foundation of the Late Rev. John Bampton, M.A., Canon of Salisbury*. Oxford: J. H. Parker.

Paley, W. (1802). *Natural Theology; or, Evidences of the Existence and Attributes of the Deity, Collected from the Appearances of Nature*. London: R. Faulder.

Prout, W. (1834). *Chemistry, Meteorology, and the Function of Digestion, Considered with Reference to Natural Theology*. London: William Pickering.

Roget, P. M. (1834). *Animal and Vegetable Physiology, Considered with Reference to Natural Theology*. 2 vols. London: William Pickering.

Rose, H. J. (1826). *The Tendency of Prevalent Opinions about Knowledge Considered: A Sermon, Preached before the University of Cambridge on Commencement Sunday, July 2, 1826*. Cambridge: J. Deighton & Sons and London: C. & J. Rivington.

Secord, J. A. (2000). *Victorian Sensation: The Extraordinary Publication, Reception, and Secret Authorship of Vestiges of the Natural History of Creation*. Chicago & London: University of Chicago Press.

—. 2014. *Visions of Science: Books and Readers at the Dawn of the Victorian Age*. Oxford: Oxford University Press.

Todhunter, I. (1876). *William Whewell, D.D., Master of Trinity College, Cambridge: An Account of His Writings, with Selections from His Literary and Scientific Correspondence*. 2 vols. London: Macmillan & Co.

Topham, J. R. (1998). "Beyond the 'Common Context': The Production and Reading of the Bridgewater Treatises." *Isis* 89: 233–62.

—. In Preparation. *Reading the Book of Nature: Science, Religion, and the Culture of Print in the Age of Reform*.

Whewell, W. (1833). *Astronomy and General Physics, Considered with Reference to Natural Theology*. London: William Pickering.

—. (1864). *Astronomy and General Physics, Considered with Reference to Natural Theology*. 7th edn. Cambridge: Deighton, Bell & Co. & London: Bell & Daldy.

CHAPTER FIVE

HOW COULD LAWS MAKE THINGS HAPPEN?

NANCY CARTWRIGHT

The question I address in this talk is: "How could laws make things happen?" My answer is: "They couldn't." To the extent that anything makes things happen it's not laws but agents. By that I mean agents in an extended sense – things in nature that have powers to act. So the kind of necessity involved in making something happen is a this-worldy necessity, not one dependent on distant laws or other possible worlds. In terms of the widely-used distinction between "top-down" and "bottom-up" influences, I shall reiterate the view that I have long argued in favour of –a bottom-up view of whatever necessity there is in nature.

Regularities in Nature

My focus will be on what I call "rough and ready regularities". There are millions of these that we rely on throughout our everyday life and in our scientific, technological and engineering practices. For instance:

- Bring up a child in the way he shall go, and when he is grown he shall not depart from it, *or...*
- Rising levels of inflation cause lowering levels of unemployment, *or ...*
- The planets circulate the sun in elliptical orbits.

It's regularities of this kind that I want to discuss. Some are rougher than others and most you wouldn't want to rely on entirely. But we do make considerable use of them in everyday life, in science and in technology. In fact, we couldn't manage without them.

Why a focus on rough and ready regularities? I started with: "How do laws make things happen?" but, more carefully, my question is: "Why do rough and ready regularities hold?" Note that I am not interested in how we explain these regularities but in what in nature makes them happen.

Is there something that makes these rough and ready regularities hold, when and where they do? I focus on these, rather than on arbitrary events, because for these we have good reason to think we *can* explain them. Though I'm not concerned with the details of any particular explanation, the *fact of explicability* remains crucial; it is the reason for supposing there *is* something that makes them hold – the very thing that explains their holding. I have a longstanding, rather down-home, view that we should avoid being committed to the idea that everything can be explained, or that everything that happens, happens for a reason, or that everything that happens has a cause. So far as I can see, the world looks very much as though some things simply happen: they occur "by hap". I want to at least allow that, because I think it's the best fit with our experience of the world, and I do not see anything like sufficient justification for the huge philosophical leap of assuming that everything that happens, happens for a reason.

Nonetheless, the point about rough and ready regularities is that we very often *can* explain them. If we can explain them, it looks as if there is something that makes them happen – it's not pure hap that they occur – and because we can explain them we think we know what that something is. That's the reason that I want to focus on these special cases. My strategy is to look at our explanations of rough and ready regularities to try and unearth the commonalities of what makes them happen, and to look to see whether or not laws are involved. In fact I'm not going to argue here that laws are not involved since I've spent a lot of time arguing elsewhere that laws aren't up to the job. I'm going to allude to that and explain it a bit, but I'm not going to argue for it. What I am going to do is to look at what makes things happen if there aren't laws.

***

I start with the observation that we have rough and ready regularities like "rising inflation causes lowering levels of unemployment", for many of which we have explanations – in this case, for example, the "rational expectations" models on offer from Chicago School economics. Philosophers have two quite different stories to tell about these explanations, two philosophical reconstructions of what's going on in them. We are not, though, in the philosophy-of-science community, disputing about the explanations themselves. We look at the very same object – what someone's done in a lecture or a journal article or a blueprint – and then try to describe it, what it's like and how it does its job of explaining, and we come up with two quite different accounts. The first is that the explanation involves fundamental laws; the second, that it involves mechanisms.

Explanation in Terms of Laws

I start with the fundamental laws story, which goes hand-in-hand with the deductive-nomological (D-N) model of explanation.[1] We are interested in what's going on in nature; we're going to look to our explanations for clues to that. Let's look at a really simple example of the D-N model. The planets circulate the sun in elliptical orbits – a rough and ready regularity. The regularity is described by Kepler's equation, which is explained by Newton's fundamental laws.

This explanation is deductive in that Kepler's equation for the elliptical orbit is deduced from Newton's, and I've no problem with the deduction bit: on both the law and the mechanism account the reconstruction has a deduction in it. But on the law account, the elliptical orbit is supposed to be deduced from Newton's fundamental laws. The D-N story has it that we write down Newton's laws, we convert the coordinates and do a little bit of mathematics and we deduce a description of the elliptical orbit. But what is meant here by "law", in the expression "Newton's laws"? This is the occasion when we're going to talk about what these D-N philosophers generally mean by "law".

Look at **F=ma**, to use a very simple, school example. **F=ma** is an equation; it states a relationship between quantities. It says that whenever particular values of **F** and **m** and **a** occur, they satisfy this equality. If we do indeed live in a Newtonian world, then whenever there's a force of size **F** on an object of a mass **m**, the acceleration is equal to **F/m**. That's a statement of association: **F** equals **m** times **a**. To be well formed it needs a quantifier in front. Most people assume it should be a truly universal quantifier, but it could have a more limited scope.

That's the claim about what's true in the world: that these quantities are always associated, or that they're always associated within some region and under some bounds ... or something like that. The equation records, we imagine, this true association between the quantities. But also, on the D-N story, the association doesn't just hold, it holds by *necessity*. That's why it's called the "deductive-*nomological*" model of explanation. The idea is that the thing you're deriving the elliptical orbit from is, itself, an association that holds by necessity, in some sense.

[1] For more on my worries about the concept of "law" and on limitations upon the scope of even high physics claims, see my *The Dappled World* (1999).

I'd like to point out two problems with this account. The first is the one I'm going to develop throughout most of the talk: the D-N model of explanation is in the "formal" mode, and I wanted to ask about the "material" mode. This is a distinction made by the famous and very deep philosopher, Rudolph Carnap. To talk about something in the formal mode is to talk about our representations of it. The D-N model of explanation is a model that talks about our representation of relations: the equations we write down and what can be deduced from them. Now we hope that our equations and our scientific theories represent true relationships: we write down relationships in our scientific theories, hoping they're the right ones, and from them the elliptical orbit can be deduced. Deduction again is a relationship between representations, it's not a relationship between things in the world. You can deduce conclusions from premises, you can't deduce happenings from other happenings.

To talk in the material mode, on the other hand, is to talk about what happens in nature. This was important to Carnap and the Positivists. They thought that it was a mistake to talk about relations like necessitation in the world. They couldn't make any sense of it. They thought we should instead reconstruct much of our talk – about causation for example or about laws – as not about the world, as the talk seems to be, but as about our representational schemes. Instead of saying "The laws represented in Newton's equations necessitate the regularity described in Kepler's equation", we should instead say simply, "Kepler's equation is deducible from Newton's equations."

Necessity

That didn't last very long. Nowadays, and for a very long time previously, people have thought, "There has to be something behind this deducibility. What's going on in the world that mirrors the deduction?" Both the premises and the conclusions in the deduction are supposed to describe facts. How do the facts described in the premises make the facts described in the conclusions true? That's what I was worried about when I wanted to know, "How do laws make things happen?" If you follow the fundamental law account, the story is that the equations represent not just relations between quantities but . It's not just that every time there is a force and a mass, the acceleration happens to be **F/m**. Rather it's that there's something necessary about that relationship. This necessity is supposed to provide the answer to my question: somehow the necessity of the laws gives them the special ability to make things happen.

How does that work? What kind of necessity is it that the regularities described in fundamental laws have, that gives them the ability to ensure that other regularities obtain? How can the fact that **F/m** necessitates the value of **a** make the regularity described in Kepler's law hold? My desideratum for an answer, to the extent that I think it makes sense to talk about necessity, is that the ability we suppose is bestowed should be true just to what we know the world is like, and not to further speculation about what it *might* be like. So it should supply the ability to ensure rough and ready regularities – which we know to obtain in the world– but it should not be moulded to be stronger than we need for that job. That means that the necessity involved should not do too much: it should be conditional, it should be partial and it should allow new things to happen.

My second desideratum is that, if there is any necessity, its source should be in this world. There are other-worldly sources that are commonly cited. I think only two are plausible, and one of these is God. This is a kind of Occasionalism: the source of the necessity of the relations between force and mass and acceleration is that, whenever God sees a force acting on a mass, He ensures that the acceleration is what it's supposed to be. That would be the Occasionalist sense of calling the relationship between **F**, **m** and **a** "necessary".

Most of my contemporary philosophical colleagues don't like that account because it appeals to the supernatural. They have another, which they seem to be convinced does not. It is that the source of the necessity for what happens in this world is what happens in other possible worlds. The idea is that, if something is necessary, then there are lot of counterfactuals true: "**F=ma** of necessity" means that, if the force operating on that mass were different, the acceleration of that mass would be different. How do we understand that according to this story? Look to another world, where the force is different; there we see whether the acceleration is different as well. If it is, then **F=ma** is necessary in our world. So the force of the necessity in this world depends on how things are behaving in other possible worlds. But what have possible worlds got over God? Even if they would be superior, I don't think there *are* any other possible worlds. *And* I don't see how, even if there are other possible worlds, what's happening in them can bestow the ability to make rough and ready regularities hold in ours.

Whether there is God or other possible worlds, I'm concerned to find some this-worldly account of what is going on. I want a story about what it is that the *explanans*[2] describes, that makes clear how it makes the rough

---

[2] *Explanans* = that which explains; *explanandum* = what is to be explained. *Ed.*

and ready regularity to be explained obtain, and I want the source of its ability to do so to be located in this world. I think we can have just that. It is mechanisms, not laws, that ensure that rough and regularities hold; and the ability to produce rough and ready regularities rests in the powers to act that the parts of the mechanism have, by virtue of their own features and their arrangement together in the mechanism. I'm happy with the explanations we give of Kepler's equation. What I'm not happy with is the deductive-nomological account of what's going on in that explanation.

Mechanisms

Let's start with a brief discussion of powers in science, using the example of forces. Einstein & Infeldt, in a little book on *The Evolution of Physics* ([1938] 1961), explain, "Forces cause motions." I think that's true. "Force" is a power word. Systems that "exert forces" cause accelerations: they have the power to make things accelerate. What then is the source of the necessity (to the extent that there is any) of **F=ma?** If you want to have an account in terms of necessity, the association has to hold because that's what the force makes happen.[3]

Let me comment on how the story about powers connects with what Michela Massimi said earlier[4]. She said that, for Kant, laws govern nature by necessitating the way natural powers act. I don't think that's the case: Massimi provides convincing arguments that it's a right reading of the early Kant, but it isn't true of the world. Taking the bottom-up view, I think the equations simply describe the consequences of the actions of natural powers. So forces have the power to accelerate masses, and if they do – if they exercise their power – then you get an acceleration that's equal to **F/m**. The equation simply describes a consequence of the action of the powers, and you can back-read what the power is from what it does. The powers act as they do, not because they are necessitated to do so by something else, like laws: powers act as they do because of what they are. If it didn't cause an acceleration, it wouldn't be a force. So part of the job that we do in physics is to learn what the powers in nature are. We also do that in biology, and we do it in engineering; I think we also do it in economics, in psychology and in sociology.

---

[3] The fact that there are forces in the world, *there's* something that this story doesn't tell. So it's not as if God is thereby excluded: He's not doing the Occasionalist thing, but we could invoke God to account for why there are these powers in the world, rather than some other powers.

[4] See Introduction, Pt 2, and Massimi (2014).

Turn now to mechanisms. It's not the laws which make things happen, I shall argue – it's mechanisms. I used to call mechanisms "nomological machines". That was a bit of a mistake. I did it because I wanted to note that these mechanisms ("machines") give rise to the rough and ready regularities that I called "laws". I wouldn't any longer call rough and ready regularities "laws": I think I did it to make people angry! Laws were supposed to be these big, grand kinds of thing represented by the equations of fundamental physics – the general theory of relativity or quantum field theory – and I wanted to point out that some rough and ready regularities were both better evidenced and more reliable than these grand "laws". But I tend not to use the term "nomological machines" much any more because I try not to call anything a law since the language seems stuck to the "necessary regularity" story of what our basic equations represent. Also, nomological machines are all the rage in philosophy just now, and my colleagues all call them "mechanisms". So that is the language I shall adopt here.

What is a mechanism? Mechanisms, as I shall use the term, are arrangements of features with powers to act such that, when these powers act together and unimpeded, they give rise to a regularity.

One example is the planetary system, in which the sun, by virtue of having a mass, has the power to attract the planets circulating round it with $\mathbf{GmM/r^2}$ strength. Or the toaster. I plug it in, put the bread in, turn the electricity on, push the lever, and get my toast. That's a nice rough and ready regularity too. In both cases we have parts that have powers in a fixed arrangement, and when those parts act together, an instance ofthe rough and ready regularity ensues. So that's my story about what makes the rough and ready regularities obtain.

My favourite mechanism is a Rube Goldberg pencil sharpener, which you can see a proper picture of in Cartwright & Hardie (2012, Fig. II.6): it is schematically diagrammed here in Fig. 5.1 (p. (101)). Because I own a mechanism like this I sharpen my pencils by flying a kite. That's a rough and ready regularity. The Rube Goldberg device is a mechanism: an arrangement of parts with powers that, acting in consort, give rise to the regularity that flying kites sharpens pencils. That's just the same as that the sun, attracting the earth, gives rise to the rough and ready regularity described by Kepler's equation, or my toaster giving rise to the regularity that if you press the lever, put the bread in, you get breakfast. Just so you don't think the parts have to be mechanical in order to constitute a mechanism, note that the nasturtium seed is a mechanism too. It gives rise to the rough and ready regularity that if you put the seed in the ground with water, air, warmth and light you get a nasturtium seedling.

Probably the most widely recognised advocates of mechanisms – the first generation – are me, Bill Bechtel (Bechtel & Abrahamson, 2005), "MDC" (Macharmer, Darden, and Craver, 2000) and Stuart Glennan (2002). This is now a fairly popular philosophic movement, namely to start with mechanisms in order to explain a lot of the regularities (in particular the biological regularities) that we observe in the world around us.

Despite the rising popularity of this approach, there's a difficulty. I'm supposed to be telling you what makes rough and ready regularities happen. But there's the same problem with mechanisms as with fundamental laws: how can *mechanism*s make regularities happen? I can write down a description of a mechanism and can deduce that a particular regularity will obtain when it operates, but that's in the formal mode. What's going on in the world? What is the relationship between the mechanism and the rough and ready regularity such that we can say that the mechanism makes the regularity obtain?

The Abstract and the Concrete

To explain what I think the relationship is, I want to go back to an easy answer to the same question for fundamental laws. I'm afraid I cheated before when I said this was a real problem for fundamental laws, because my undergraduate teacher, Adolf Grunbaum, taught me the answer, and I think he's right if you believe in fundamental laws: it's a matter of the relationship between the abstract and the concrete. What Grunbaum said is: "The elliptical orbit of Kepler's laws is just what it is for Newton's laws to be true in that arrangement. There's nothing more to it than that." If it's a planet, moving like this, if it's a canon ball moving like that, each of these is what it is for Newton's laws to be true in their particular arrangements. So Adolf Grunbaum claimed that there's no problem about how Newton's laws make Kepler's equation true, because the behaviour described in Kepler's equation just *is what it is* for Newton's laws to be true in that arrangement. Just by being true themselves, Newton's laws make Kepler's equation true.

My answer to the question about what happens for mechanisms is just the same. It's a matter of the abstract and the concrete. The abstract is the exercising of the general powers the parts have – for the planetary system, the gravitational power of the sun to attract and the power of the planets to be attracted; what that amounts to in the concrete, in the arrangement of the planetary system, is the planets moving in an elliptical orbit. The point is that the arrangement allows the general powers of the various parts of the mechanism to be exerted in consort, which they otherwise could not be. So the arrangement makes all the difference.

I got these ideas from two sources. First, my co-author John Pemberton, who has spent a lot of time working in the mechanism literature. Other mechanists, he claims, don't pay enough attention to the arrangements: they talk about parts and their activities, but they don't talk much about their arrangements. The second source is my own work on the abstract and the concrete, which I have illustrated with the example of parables or fables. There's the moral of the parable or fable, which is usually a very abstract idea, and then there's the parable or fable itself, which makes the moral concrete in a given situation. Or, we teach our children that they should not hurt other people's feelings – that's an abstract instruction. What they have to learn, occasion to occasion, is what constitutes not hurting people's feelings on those particular occasions.

One of my favourite fables is by Lessing, the German Enlightenment playwright, who argued for just this interpretation of the relation between fable and moral using this as an example: "The grouse is eaten by the marten, the marten is eaten by the fox, and tooth of the wolf gets the fox." The moral is, the weaker are prey to the stronger. If you're a grouse in the vicinity of a marten, being a grouse is what it is to be weaker and being eaten by the marten is what it is to be prey to the stronger.

Now let me lay out how rough and ready regularities are made to happen on the mechanistic account. Powers have canonical, though defeasible[5] activities associated with them. So a mass has the gravitational power; it pulls other masses towards itself. Sometimes, due to the other features of the arrangement in the mechanism, a feature will have a power it otherwise would not. My flying kite is a clear example. If you look at the picture, the first step is that flying the kite opens the door of a little cage and lets some moths out. Flying kites though does not generally have the power to release moths. Yet it does have that power in my Rube Goldberg machine, and that's due to the arrangement: due to the other features in the mechanism, the kite-flying turns out to have a new power – to open the little door. The arrangement of the parts allows the canonical activities of these powers to combine when they otherwise wouldn't. The result is that the individual parts acquire powers they wouldn't have otherwise. They combine in this simple case by sequencing: by one power acting after another. The rough and ready causal regularity, that kite flying sharpens pencils, holds on account of these combined causal activities, activities that wouldn't otherwise combine. That's the punch-line – the

---

[5] *Defeasible* = capable of being counter-argued by adduction of new facts. *Ed.*

kite flying sharpening the pencil just *is* what it is for all the activities of the powers of the parts to occur and combine, in the way fixed by the arrangement of the parts in the mechanism.

KITE	SHIRT	IRON	OPOSSUM.	WOODPECKER
pulls on	releasing	which burns	It jumps into	which chews
STRING 1	STRING 2	PANTS	BASKET	PENCIL
which lifts	which drops	causing	pulling on	exposing
DOOR	BOOT	SMOKE	STRING 3	LEAD
releasing	onto	which enters	opening	with which
MOTHS	SWITCH	CAVITY	CAGE	NANCY
which eat ..	heating ..	disturbing ..	releasing ..	can write!

Fig 5-1: Schematic diagram of Rube Goldberg's Kite-activated Pencil Sharpener. Read each column top-bottom, then go to top of next column to the right

Consider the pencil sharpener again. Look at the first step: flying the kite opens the door. That's because of the double pulley. Power 1, the first activity that's exerted, is that pulling up on the input end of a double pulley system raises weights at the output end. (This isn't proper physics – it isn't even proper Archimedian physics – but it's a lay-language description that will serve.) Raising a weight at the output end of the pulley rope when the rope is pulled on the input end is the canonical activity of the power bestowed by being a pulley. Now consider the door-opening – it releases the moths. We have here another activity of another power: breaching a closed container releases a fluid inside. There is then a second power involved, and the breaching is the activity of that power. These two activities occur one after the other: pulling up on the input end of a double pulley raising a weight at the output end, followed by breaching a closed container releasing a fluid inside.

The point is, though, that normally these two activities don't compose. We have: a) pulling, **P**, causes raising, **R,** and b) breaching, **B,** causes fluid flow, **F**. What we want is that flying the kite, which is the concretization in this setting of pulling on the pulley, causes the fluid to be released. How do we get that? We have the separate powers, the pulley power and the power of a breached container to let its fluid out, but how do the two manage to connect **P** and **F**? They don't compose as they are: **P** won't directly cause **F**. But the arrangement here ensures that they do compose. That's the trick. The arrangement allows the two powers to compose: In the concrete, **R** (raising the door on the moth cage) and **B** (breaching the container) are one and the same event. That one event is a concretization of two distinct abstractions: because of the arrangement, raising the weight at the end of a pulley *is* opening the door of the moth-cage.

So there is a very simple story about how mechanisms give rise to rough and ready regularities. It's the same thing that Adolf Grunbaum taught me when I was an undergraduate at Pittsburgh: it's simply the relationship of the abstract with the concrete. But it is the arrangement that allows the abstract activities of the parts of the mechanism to combine appropriately to constitute the activity of kite-flying sharpening pencils. The rough and ready regularity (like: flying the kite causes pencils to be sharpened) just is what is for all the parts to exert their powers together in the arrangement of the mechanism. Rough and ready regularities are indeed made to happen, but they are made to happen not by laws, but by mechanisms.

Discussion

Questioner 1: I thought you were saying that the arrangement of a mechanism allows an abstract process to be instantiated in the concrete. But I didn't find anything more concrete in the concatenation of the Rube Goldberg bits than they provide individually.

Answer: The arrangement does two things. First, it's only on account of the arrangement that the raising of the door by flying the kite *is* the concretization of pulling on one end of a pulley and so raising a weight at the other end. It's because of the arrangement – that the kite is on a string, the string is relatively unbreakable, it is relatively friction-free, it goes across the double pulley, it is tied very tightly to the little door – that pulling on the end of the pulley raises the little door. That's the first thing it does. If it weren't for that arrangement, the flying of the kite wouldn't be a concretization of the power of the pulley.

Second, I've got these two powers, the power of the pulley and the power of the breached container to release a fluid from the inside. Normally, having these two doesn't help: put those two powers in one bag and they can't do anything together. Exercise the power of the pulley and something goes up; exercise the power of the breaching of a container and something escapes. They aren't linked. But the arrangement of the mechanism has allowed those two powers to concatenate by this neat trick that the concretization of the effect of the first one is at the same time the concretization of the cause of the second one. So now, at the concrete level, **A** causes **B, B** causes **C**, therefore **A** causes **C**. That wouldn't happen if it weren't for the arrangement.

Q: May I ask a rider? Where's the arrangement in Kepler that allows us to see where mechanisms can apply?

A: That's fairly simple. We don't have a sequenced concatenation: we have just the power of the sun to attract, the powers of the other massive object to be attracted, and that little bit of extra force you need to get the acceleration right. Concatenation of powers – getting them to work together – is notoriously easy with forces; as soon as you have different forces exerted on the same point they just concatenate, by vector addition. So all one has to do with forces is to get an arrangement that allows them to be exerted together.

Those are the simple cases of concatenation that you get all the time. Consider two electrons. They're attracting each other because of their masses, they're repelling each other because of their charges; all you have to do, to get these two powers to act together, is to get the two electrons anywhere without a shield between them, and that's the arrangement. It's easy to get the abstract powers exerted together, because you just have to get the particles together any old way.

Questioner 2: Does concatenation underpin the concept of emergence?

Answer: I think it can underpin at least one sense of emergence. We create new things all the time, by putting powers together in very special and new ways. That may be all there is to a lot of the cases we call emergent. I'm not prepared yet to say it *is* all there is to it, but I am looking at many cases again in my research for the Durham Emergence project, to see how far this idea will go.

Questioner 3: How do you define where the mechanism begins and ends? You're flying your kite, but for that you need a windy day, and for wind you need Coriolis forces in the atmosphere, and to get those you need the rotation of the earth, and so on ….

Answer: That doesn't bother me. It seems young folks working in philosophy on mechanisms are trying to come up with identity criteria: when are two variants the same mechanism? But I don't see why this is an interesting question for my project. If the point is to find the mechanism that gives rise to a particular rough and ready regularity, the answer is trivial: It takes whatever it takes. For the pencil sharpener at least, nature does not need the wind as part of the mechanism since it is a mechanism intended to generate a regularity about what happens *when* the kite flies, not one intended to fix *whether* the kite flies. But I also think you will always need a "non-interference" clause if you want to state the conditions for the regularity to obtain: If a mechanism of this design operates unimpeded, this regularity will obtain, where what counts as unimpeded is fixed by the facts about what can and cannot interfere with the exercise of all the powers involved. (And it is part of the ontology of powers, I maintain, that there is a fact of the matter about whether a particular happening on a particular occasion is or is not an impedance.)

Questioner 4: I've been trying to put your arguments into two contexts. The first is that of theology, which I prefer to see as a form of enquiry, not as a revelatory system. I quite like, and have tried before this to work with, the Carnap distinction between formal structure and mechanisms. The idea of arrangement struck me as interesting: do you take it into account in that sort of framework? But I then thought, "How do you apply this in an area which you've looked at a great deal, economics?" If you take something like trade cycles, how would you relate the formal structure to the mechanisms to the arrangements, in that context?

Answer: I'm going to do the same thing in relation to economics that I suggested as a strategy for physics. Let's look at our explanations of certain putative rough and ready regularities to see what's going on in the world of which these are concretizations. Let me tackle the example I started with, the inflation/unemployment one. We have rational expectations models that seek to account for the short-term regularity that rising inflation causes lowering of unemployment. The model has in it rational agents: entrepreneurs. The entrepreneurs see prices rising but do not recognize that the price rise is a general inflationary price rise; they mistake it for a price rise in their industry, so they expect they will make profits if they produce more, so they open new jobs and unemployment goes down. Then there's the follow-up story, that Robert Lucas and other Chicago School economists maintain: if the government tries to use this as a lever, that story won't work anymore, because once the government manipulates inflation, it's public, and the entrepreneurs know it's a general rise in prices and not a price-rise in their area.

In the model, the entrepreneurs only care about profit. They have either full or limited information.... And so forth. Those are the kinds of powers ascribed to the agents. We write these down, we write down the formula for the expected utility, do the derivation, and it seems to me that this doesn't look like a derivation from general laws. You could *pretend* it's a deduction from general laws, because you could say it's a general law that people always act to maximize their expected utility, and in this case the expected utility of the entrepreneurs is profit. But I think it would be a mistake to consider that it's functioning as a general law. Rather, this is one of the powers that agents have: to act to maximize their expected utility. And that power often quarrels with other powers they have, such as to care about living in a decent society, to care about their reputation, etc. ..... But maybe that didn't answer your question?

Q: That's certainly very helpful, but what if the entrepreneur isn't rational?

A: Well, rationality in my book has to do with reasoning and thinking and having good arguments and good reasons and so forth, whereas what's called "rational" in economics is that you act to maximize your expected utility. So I would call these "greed" or "self-centred" models. And when the entrpreneuer isn't greedy in this sense, there the rough and ready regularity relating inflation and unemployment will probably not hold.

Questioner 5: I was thinking about the outcomes of the mechanisms, and their value. For instance, your pencil sharpener could set fire to your house instead of sharpening your pencil, and that would be a worse outcome for you – but that's your perspective, that's abstract.

Answer: The notion I am employing is a relational one, "more/less abstract than". So describing the situation as "the weaker is the prey to the stronger" is more abstract than "the grouse is the prey to the marten". I don't quite know how to answer the question then, because I need a comparison.

Q: I suppose I'm thinking about the fact that the mechanisms have an outcome, and the outcome is in our looking at it.

A: I agree.

Q: But all other outcomes would be equally significant in terms of mechanism.

A: That's right. This mechanism gives rise to lots of other rough and ready regularities.

Q: I suppose really what I'm driving at is that, 20 years ago I had a debate with John Polkinghorne about the Anthropic Principle, and he looked at me and said, "You must be a biologist!" There could be all sorts of outcomes, and they're all equally significant.

A: I agree with that, but pencil sharpening happens to be the rough and ready regularity I was interested in. Of course, I would be very interested if, in general, there was a regularity that when I ran the pencil sharpener my house burned down. But I don't see any grounds for the idea that this particular mechanism, operating unimpeded, would give rise to a regularity involving my house burning. I do agree, though, that there are lots of other regularities that it gives rise to that we don't bother to note or think about or invite explanations for.

Quetioner 6: This is a question about the arrangement of the powers. I think, in the last slide, you suggested that it was a temporal arrangement?

Answer: In this case, yes.

Q: Yes. So could there be other kinds of arrangement, like the spatial arrangements that seem to apply to molecules in organic chemistry?

A: The arrangement in this case was very complicated. The picture I showed you was the spatial arrangement of the parts. What it allowed was a temporal concatenation: how it gave rise to the regularity we're interested in was that it allowed for a temporal concatenation. But in the Kepler example the arrangement of co-presence directly allows for a concatenation of the forces. All you have to do is get the forces to be in the same place at once, and they all act at once as vector **F**.

So there certainly are other ways in which you can do concatenations besides temporal. You can, for instance, get a more elaborate description of the Goldberg example that has vector addition in it, because pulling on a pulley doesn't always raise something at the other end. Yet if it's successful it'll cause a force to be exerted, an upward force at the other end, and then you have to think about the weight of the door and the friction …. Because the little door slides on some runners, there will be a frictional force; the arrangement allows this frictional force, the upward pull, and the downward pull of gravity all to be exerted at once. So the arrangement doesn't just allow for temporal concatenation, it allows for other concatenations to occur too.

The point, as I now see it – though I've only worked through a few examples – is that clever people figure out how to take some general knowledge they have about the powers of materials, and put them together so as to do things we never previously dreamed could be done. It's not that they make up the new garment out of whole new cloth: they use knowledge they had before. But how can I best represent what they're doing? They're certainly not generally writing down a bunch of laws and deducing things, but they are somehow or other figuring out how to get general powers that they already know about to produce new effects. For instance, they know about the elasticities of various media, or that certain substances will superconduct if you cool them below certain temperatures – they use that kind of knowledge about what things have the powers to do, and then deploy it in a really clever way by making a design that allows those powers to work together to produce a novel result that could not emerge if the same powers were just stuck together in an arbitrary way.

Questioner 7: I was wondering if you were trying to abolish Occam's razor?

Answer: I don't believe in Occam's Razor, but I wonder where you saw that here?

Q: Well, my impression is that you're wanting to replace some beautifully simple physical conceptions (admittedly mysterious in origin) with a lot of complicated things.

A: I hope not! We know, I suppose, a lot about superconducting materials or about masses attracting or electrons repelling each other or about space-time curvature causing precession of gyroscopes …. What I am doing is claiming that the standard story, about what it is that we know when we know those things, is mistaken. It isn't that we know general laws but that we know about the powers of things: interpreting our simple conceptions in that way makes a more realistic picture of the way the world is.

To take that point further, I'm far from convinced that there *are* any general laws. Admittedly there seems to be a general truth about how forces combine. *That* doesn't seem to be the exertion of a power, it's a story about how powers combine. This looks like a general law, but otherwise I don't see any general laws. I see facts: that there are certain kinds of powers in the world.

This is all reconstruction, of course. If you look in a physics text, or at a physics journal article, or a diagram of how to design, say, a laser, it won't say, "Here are some laws", but nor will it spell out a set of powers. Instead there will be equations, and bits of information, and approximations, and so on. So whether we talk about general laws or about powers, either way we're making a philosophical reconstruction. (And it's as much a philosophical reconstruction when scientists do it as when philosophers do it.) So yes, there's a sense in which we're making things complicated by asking what's really going on in this journal article where something's derived from something else. But I don't think that this has much to do with my disbelief in Occam's razor.

Questioner 9: I'm very happy that you are getting away from all these scientific laws, and reducing them to things that actually happen. But can you reduce *that* to there being nothing that is absolute? Suppose I'm trying to reduce matter to its smallest bit, maybe a string or something. Will I not eventually have to come to something or other which genuinely and absolutely physically *is*, and necessarily is? Wouldn't that something, and the descriptions of its behavior, be laws – a base from which everything is built?

Answer: Well, it could all be built as you describe. But, once again, there would be the two philosophical stories. On the first there would be some fundamental laws, laws that say, "There are these kinds of particles, and they satisfy this equation, which is necessary in some sense." You could have that story, and then try and tell how other things happen on account of those laws being true. On the alternative, there would be fundamental particles that have certain powers, but that doesn't yield anything about universal behaviour: when they're not in the right circumstances, there may be nothing fixed about what will happen.... What occurs, occurs by hap. That would be the difference between the law story and the power story. Either way, you can start out with the same physics, the same fundamental particles, the same claims about them, and you can reconstruct these claims as fundamental laws that are there all the time, or you can reconstruct them as powers that get exerted in certain circumstances. To which we can add that we don't know what happens (or maybe nothing's dictated to happen) outside those circumstances. They're two different pictures. But it seems to me that one of them, the law story, is over-reaching. By contrast the power story may sound like a lot of heavy metaphysics, but it doesn't say more than we know.

Do you know the story that Otto Neurath talked about? If you drop a 1000 mark note from a high building in a city square, it goes here, there and everywhere, unpredictably. If you believe the story about laws, and you're a real die-hard, you'll say that we know the bill is pushed by the wind, that gravity also acts on it, and both these must be represented as forces which have to add vectorially. Finally, if you know the mass of the bill, you will contend that its acceleration at any one moment can be calculated from this summated force. What I claim is that we don't know, one way or the other, whether that story is true. When you can represent everything that acts on the bill as a force in a proper way – not doing it *ad hoc*, by back-reading what the force has to be to produce the observed acceleration – when you can actually assign the forces in a legitimate way, the way we assign the values in $\mathbf{GmM/r^2}$, you've got a real description of the situation; then you can justifiably claim that it's only operated on by forces. But in the meantime, to claim that everything that can cause a motion is a force is just reaching beyond what we know.

Contrast this with the "powers" ontology. If the story about everything being represented by a force is true, you can fit that into the powers ontology, but if it isn't true, *that* can be fitted equally well. We don't commit ourselves to there really having to be a law which governs such situations in a systematic, universal way, the same law governing all motions, everywhere, all the time.

Questioner 10: I'm still grappling with what feels like a semantic trick. Excuse me if I've missed the point, but can what you said be reduced to something like: "How do rough and ready regularities occur, how do we explain them? Well, we say, things are sometimes arranged so that something happens which is the rough and ready regularity."

Answer: That's right, as the starting point. But, in order to see how the arrangement works, I think you're going to have to have powers. When we do a derivation, the question is whether or not we can reconstruct it, thinking that the premises are all statements of necessary, regular association. But, whether you do fundamental laws or you do mechanisms, I think the right answer is the Grunbaum answer, which is essentially the one you stated. However I think the Grunbaum answer is much more attractive when deployed in relation to mechanisms. The arrangement allows that the sharpening of the pencil by the kite flying just *is* the operation of the mechanism.

Q: That seems to me to re-insert words, like 'necessary' and 'allow', which I thought we were trying to get away from. It seems as though the simple constitution answer enables us to cut to almost a tautology: it enables us to get rid of laws, but …

A: I believe I did argue that things are *made to happen*, so there is necessity of at least this kind. The mass of the sun makes the earth move towards it; that's what happens when the mass of the sun exercises its power to attract another mass unimpeded. More generally what makes rough and ready regularities obtain is the exercising, in consort, of the powers of the parts of the associated mechanism. The exercising of the powers make the effects at the tail end of the regularity happen, when the events at the front end of the regularity trigger the powers to act. Producing the effects at the tail end of the regularity just *is what it is* for all those powers to produce their effects in consort.

What we don't have here, though, is *unconditional* necessity. The arrangement matters. The arrangements I've been talking about ensure that powers act together in a systematic way. But I can imagine that there are many arrangements where the powers just have a tug of war, with one winning sometimes, another at other times, and no systematicity to it. We at least have no compelling reason against this. So I think I have satisfied my own desideratum that the kind of necessity I have urged is partial, it is conditional and it allows new things to happen.

Q: These are always models: rough and ready regularities are models in the sense that we can't say that they actually *are* the case. We've observed lots of things that look the same to us, and so we make a model and say that when inflation goes up, this happens to the economy, or whatever.

A: Well, I think it *actually happens* a lot of the time that when I knock a cup it falls, and that if you bring up a child in the way it should go when it is grown it will not depart from it ….. It doesn't happen all the time, but it surely happens a lot of the time. It also happens a lot of the time that the weaker are a prey to the stronger. These are real occurrences, not models.

Questioner 11: I was wondering if you meant by "mechanisms" the same things as Mumford (2004), Ellis (2001, 2008) and Bird (2007) seem to think of as "dispositions", and whether you'd class yourself as a dispositional essentialist?

Answer: I think they mean the same thing by mechanisms as I meant by "nomological machines". "Mechanisms" caught on about 15 years after I was writing about nomological machines.

Q: I wanted to find out whether you thought that objects could have different mechanisms, or whether you believed in the "dispositional/categorical" distinction?

A: One thing Mumford thinks is that there's nothing but dispositions. I don't quite know how to make sense of that. I think that the kind of informal science that I understand is that there are features – for instance, masses – features I can independently identify that bring with them powers. I want to stay neutral about whether the feature of being a mass, or having a mass, the feature "massiveness", is just a collection of powers, or it's something else (maybe a categorical property) such that, whenever a system has this something else, it has the power to attract other masses. Mumford thinks there are just dispositions and that dispositions are dispositions to produce further dispositions . I find it difficult to think through how to decide whether you could have one collection of powers which, when exercised, give rise to another feature, such as faster and faster motion, which is itself just another collection of powers. That bit of metaphysics is one I don't indulge in, one way or another, because there's nothing I've seen, in the scientific uses which have driven me to believe in powers, that would decide that issue.

Questioner 12: I wonder how you react to the slightly humorous suggestion that you are in fact proposing another fundamental law – that rough and ready regularities are made to happen by mechanisms and arrangements?

Answer: If a law is a necessary association between quantities, then I'm not proposing a general law. I do think the claim you've enunciated might itself be a rough and ready regularity: i.e. that mechanisms give rise to rough and ready regularities is itself a rough and ready regularity. They don't always do so, and it might be that some rough and ready regularities just happen by hap. So I think it's not a general law, but it might be something that regularly happens – i.e. it's a rough and ready regularity!

References

Bechtel, W. & Abrahamson, A. (2005) "Explanation: A Mechanistic Alternative." *Studies in History and Philosophy of the Biological and Biomedical Sciences*, 36, 421-441.

Bird, A. (2007) *Nature's Metaphysics: Laws and Properties.* Oxford: Oxford University Press.

Cartwright, N. (1999) *The Dappled World: A Study of the Boundaries of Science.* Cambridge: Cambridge University Press .

Cartwright, N. & Hardie, J. (2012) *Evidence Based Policy: A Practical Guide to Doing It Better.* Oxford: Oxford University Press .

Einstein, A. & Infeldt, I. (1961) *The Evolution of Physics (edn 2).* Cambridge: Cambridge University Press [1st edn 1938] .

Ellis, B.D. (2001) *Scientific Essentialism.* Cambridge: Cambridge University Press.

Ibid (2008) "Essentialism and natural kinds" in S. Psillos & M. Curd (eds), *The Routledge Companion to Philosophy of Science.* London: Routledge.

Glennan, S. (2002) "Rethinking mechanistic explanation." *Philosophy of Science,* 69 (Suppl.), S342–S353 .

Machamer, P., Darden, L., & Craver, C. (2000) "Thinking about mechanisms." *Philosophy of Science*, 67, 1–25.

Massimi. M. (2014) "Prescribing laws to nature". *Kant-Studien,* 105, 491-508 .

Mumford, S. (2004) *Laws in Nature.* London: Routledge.

SECTION II:

OFFERED CONTRIBUTIONS

CHAPTER SIX

LAWFULNESS, BIOLOGICAL CONTEXTUALITY AND A THEOLOGY OF INTERDEPENDENCE

FRASER WATTS

This paper is primarily about the application of the concept of laws of nature to complex systems, especially as found in biology. As biology becomes increasingly holistic and organismic, there is increasing emphasis on complex systems in which there are numerous , and the concept of laws of nature sits uneasily in sciences that emphasise complex systems. Such systems can be found in most sciences, but I suggest that they are a particularly conspicuous feature of the living world, and are receiving growing emphasis in contemporary biology. Though I argue that the concept of laws of nature sits increasingly uneasily with biology, the new systemic[1] biology is theologically suggestive in other ways that I will explore later in this chapter.

The "Laws of Nature" Concept

I will begin with a brief indication of why I am sceptical on general grounds about the concept of laws of nature, though these issues are explored more fully elsewhere in the volume. First, talk of laws of nature suggests a reification of the lawfulness of nature, as though invoking laws of nature explained that lawfulness rather than just being a re-description of it. If the "laws of nature" are really just generalizations about how nature normally operates, as many have suggested from David Hume onwards, the concept of "laws" is misleading, as it suggests that that there is something external to nature to which nature is having to conform. That may make sense within a theistic framework, but I suggest that it is not easy to formulate the concept of laws of nature in a coherent way without

---

[1] *Systemic* = concerned with the whole organism, or interactive systems within it. Contrast *Systematic* = concerned with the classification of organisms. *Ed.*

a general metaphysical position such as theism. It seems to be an example of an idea that has migrated across from theistic to non-theistic thinking, but which doesn't really make sense in its new secular context.

Second, talk of "laws of nature" exaggerates nature's lawfulness in a way that is potentially misleading. Even in biology, there is certainly an element of predictability to study, but that predictability is so complex that exact predictions are hard to make, as they are with the weather. The causal processes involved are so complex, and so dependent on context, that they just don't exemplify the kind of simple, linear determinism suggested by talk of laws of nature. As Prigogine famously observed (Prigogine & Stengers, 1984) there are more clouds than clocks in the world, i.e. more systems that are so complex that in practice they are unpredictable. Talks of "laws of nature" makes it sound as though the world consists mainly of predictable mechanisms like clocks. Whether or not complex processes are actually indeterminate, or just too complex to predict, is a subtle issue, though I share Polkinghorne's view that complex processes in nature are probably indeterminate (Polkinghorne, 1998). However, that is, as he admits, only a reasonable metaphysical conjecture.

Third, talk of "laws of nature" seems to imply a linear top-down process, whereas I argue that lawfulness, especially in biology, is often complex, interactive and systemic. As Peacocke (2001) emphasised, there are often both top-down and bottom-up causation, with wholes influencing parts as well as parts giving rise to wholes. We also need a way of understanding lawfulness that allows for emergentism, i.e. for genuinely new phenomena arising at higher levels that are not predictable from lower levels (Clayton, 2006).

I do not deny that nature is, in a broad sense, lawful. If it were not so, scientific enquiry as we know it would have been impossible. In addition, those who look at the natural world from a theological point of view will reflect that lawfulness is central to God's purposes for his creation. If nature were not as lawful as it is, it would not have been as fruitful as it has been, nor served God's creative purposes so well. However, it is a big leap from the recognition that nature is, in some sense, lawful to the reification of "laws of nature".

Systemic Biology

The concept of "laws" of nature works better for some aspects of the natural world than others. I suggest that it works better for simple than for complex systems and that, because the living world is so radically systemic, it is especially misplaced in understanding complex biological systems. Biology has never been an exact science and deals with probabilistic processes much more than do the physical sciences. There has recently been such a marked trend towards a recognition of the need for more systemic, organismal explanations in biology that the concept of laws of nature seems especially misplaced there. This trend is seen most clearly in epigenetics (e.g. Francis, 2012), which emphasizes the contextuality of Mendel's so-called "laws". Biological processes are so complex and systemic that the concept of "laws of nature" is not a convincing way of conceptualising the lawfulness of biological processes.

There is a strong holistic tradition in biology that goes back a long way, though the astonishing achievements of molecular biology and biochemistry eclipsed that tradition for a while. C. H. Waddington, who coined the term "epigenetics" in 1942, is an example of this rich, older tradition of more holistic biology (Waddington, 1942). Developmental biology was always a stronghold of the organismal, holistic approach, and Brian Goodwin (1994) was one of its most effective champions through what were, from the holistic point of view, the dark ages of biological reductionism.

Now the pendulum is swinging back, driven by cutting edge empirical research, with epigenetics leading the way. Denis Noble has become one of the most outspoken champions of the new systems biology (Noble, 2006). It is now clear that genes don't just exercise their effects in a deterministic way, regardless of circumstances, but that there are complex systemic factors that switch genes on or off. Mendel's so-called laws are dependent on context. Indeed, as the genome has been mapped, it has become clear that much of the DNA seems to be serving these regulative functions rather than directly determining phenotype characteristics. The old picture was of people being controlled by their genes, and Richard Dawkins' first book, *The Selfish Gene* (Dawkins, 1976) popularised that picture. In fact, the effect that our genes have on us is balanced by the effects that we, or at least our situation and circumstances, have on our genes. As Peacocke would have said, wholes influence (or constrain) parts; it is not just that parts determine wholes (Peacocke, 2001).

But it is not only epigenetics that is driving this return to holism in biology. There are also, for example, interesting implications of recent research on neural plasticity. What we are able to do depends on the structure of our brains. However, it is equally true that the structure of our brains reflects our pattern of activity. The most famous example is probably the way in which the hippocampus becomes unusually developed in London taxi drivers (Woollett & Maguire, 2011). There is also the growing field of embodied cognition, with numerous demonstrations that cognitive processes such as judgements are dependent not only on neural processes in the brain but on patterns of activity in the body as a whole (Teske, 2013). Once again, specific detail is influenced by broad context.

There is also ecology, which has increasingly shown that the prospects for one species are very dependent on the fortunes of other species with which it is inter-dependent. A classic example is how the dams that beavers build create a still-water environment that enables other animal species and an abundance of rich vegetation to flourish; beavers are thus great creators of habitat. There is an established tradition of eco-theology that has reflected on such things (e.g. Grey, 2003).

Michael Denton comments that "the design of living systems, from an organismic level right down to the level of an individual protein, is so integrated that most attempts to engineer even a relatively minor functional change are bound to necessitate a host of subtle compensatory changes" (Denton, 1998, p. 342). I agree with Denton about that, though I don't share the scepticism he expresses elsewhere about Darwinism. However, I do think we need a Darwinism that is broad and subtle, such as that of Simon Conway Morris (Morris, 2008).

What are the theological implications of these developments in more holistic biology? I suggest that the new systemic biology promises to be fruitful theologically in several ways. It steers biology away from the crude reductionism that has been the point of greatest tension between biology and theology, a reductionism that seemed rather compelling after the triumphs of biochemistry. It also suggests a more subtle view of divine action in which God's purposes are effected gradually through engagement with the complex systems of creation rather than by discrete interventions. Further, it invites us to connect the inter-dependence that is increasingly evident in nature with the inter-dependence that is assumed to be central to the nature and purposes of God. These implications will be considered in turn.

Beyond Reductionism

First, the new, more systemic biology liberates us from strong forms of reductionism, what Dennett has called "greedy" reductionism (Dennett, 1995). Reductionism proceeds through two distinct stages, starting with the assumption that higher-level phenomena are completely predictable in terms of lower-level processes. It is one of the strange features of reductionism that convincing evidence for that assumption is seldom presented. Indeed, there seem to be very few, if any, interesting human phenomena that are completely predictable. However, it is simply assumed, ahead of any actual evidence, that science will eventually deliver complete predictability, even if it has not yet done so. It is hard to see where such confidence in predictability comes from; it seems to be an act of faith, and to arise from an ideological view of the power of science rather than from the track record of science.

The new more systemic biology is incompatible with complete reductionism. A systemic approach to biological theory assumes multiple, mutually-interacting causal influences. That precludes complete reduction to a single set of low-level causal factors. For example, the systemic assumptions of epigenetics, in which genes are recognised to be influenced by organismal and contextual factors, preclude reduction to genes alone. The more complex the network of causal influences, the harder it will be to specify causal factors so completely as to allow full and complete prediction. In that sense, the conditions of determinism seem unlikely to be achieved in practice and, in as far as complete prediction is achieved at all, it will be based on multiple causal factors, so precluding reduction.

What is at stake here theologically is focused more on the further moves that are often made once complete prediction from low-level processes is achieved, or assumed. It is then often suggested that the higher-level phenomena are somehow not quite what they seem, or are just an epiphenomenon of lower-level processes. A classic example is Francis Crick's reductionist thesis about the human person. He assumes that everything about humans, including consciousness and soul, can be predicted from brain processes (Crick, 1994). The conclusion that Crick draws is that people are really just a bundle of neurons, and that the common supposition that they are more than that should be discounted.

There is also a strong strand in religious thinking that emphasizes human dignity (Soulen & Woodhead, 2006) and the significance of humans in God's purposes. That is inconsistent with strong, "nothing but" reductionist claims about the human person. Philosophical solutions have been proposed to the problem of reconciling the physicalist assumptions of science with the dignity and significance of humanity, such as non-reductive physicalism (Brown *et al*, 1998) or emergentism (Clayton, 2006). Now we have a solution to this problem that comes from scientific research itself, i.e. that causal processes are too complex and systemic for complete reductionism to work. I suggest that a scientific solution to the problem of reconciling science and religion on this point is likely to be the most convincing and enduring solution of all.

Divine Action

A second reason for religious thinkers rejecting strong reductionism is because it leaves no room for God's action in the world, or indeed for any influences other than those recognised by materialistic determinism. Because the new more systemic biology leaves scope for multiple influences of different kinds, and because it is not saying that there is not just one kind of causal process (bottom-up), it is in no position to rule anything out. It follows that it is in no position to rule out God's influence in the world. As Peacocke (2001) suggested, there is at least an analogy between whole-part influence in nature and the influence of God on creation. Because the new systemic biology is increasingly recognising top-down processes in biology, it may be more open to an analogous process by which God influences the world. At least one objection to that idea, i.e. that there are no causal influences other than bottom-up ones, is removed. One way of thinking about God is as the ultimate context for creation, the ultimate whole in relation to which/whom everything is a part.

Holistic biology also suggests the complexity of God's influence in the world. Because of the radically inter-connected nature of the living world, it is not really credible that God should act in the world by intervening at a single point to bring about a specific outcome. The fact that biology is discovering that the living world is subject to multiple, mutually-interacting influences suggests that, if God is to act in the living world, God is usually likely to do so through the confluence of multiple influences, rather than through a single, decisive intervention. Because God's creation is systemic and inter-connected, God's influence within it is likely to weave within the created order in multiple ways in order to set forward God's purposes. It seems to have been one of the unintended consequences of our inheritance from Augustine and others that we have often framed issues of divine action in terms of God acting on his creation from outside (Knox, 1993). A view of divine action that was influenced by the new more systemic biology might help us to recover the patristic view of creation being enfolded within the life of God. On that view, to ask how God can intervene in nature, as though it was something separate from him, is to start from the wrong place.

Abandonment of a simplistic view of "laws of nature" in favour of a more systemic view of multiple influences suggests that discussion about whether God might overturn the laws of nature in order to act in the world is misplaced; it presents a false and over-simplified picture of the lawfulness of creation that God might overturn. In recent centuries miracles have often been defined in terms of whether they run contrary to laws of nature, as in the investigation of Vatican into whether healing miracles have occurred at places such as Lourdes, but that is an unsatisfactory way of conceptualizing them. Historically, it is arguable that defining miracles in that way only became fashionable after nature came to be understood in terms of "laws of nature" in the seventeenth century, and that it replaced an earlier and more subtle way of understanding miracles. Seeing miracles as contrary to the laws of nature also assumes that scientific and theological accounts of exceptional events are alternatives, whereas I would argue (e.g. Watts, 2011) that they provide complementary and mutually-interacting perspectives.

Interdependence

A third respect in which the new, holistic biology is theologically constructive, is that it emphasizes interdependence between wholes and parts, and between structure and function. Similarly, ecology shows an interdependence between species in a complex ecology in which different species perform important functions for one another. This inter-dependence in creation is suggestive, theologically, of the interdependence between the different persons of the Trinity, in which each person within the threefold God is distinct from the others, but each participates in the others. Contemporary biology, especially ecology, is discovering in creation a kind of *perichoresis* (dance of love) that is an echo of the perichoresis that Christian thinkers have claimed takes place between the three persons of the Trinity. This inter-connectedness seems to be a particular feature of the living world, though the discovery of quantum non-locality suggests it may also be a feature of the quantum world too. It is arguable that inter-connectedness is one of the most important ways in which creation echoes the Godhead and is an image of God.

It is also notable that there is a striking consonance between the picture of an inter-connected world that is now coming from biology and the mystical vision of the unity of all things. It is probably fair to say that mystical experience focuses primarily on the sense of unity with the Divine, though there is also often a secondary sense of the unity of creation. The two are often linked. If the Divine is experienced as a sacred unity, with whom it is possible to be at one, everything in creation is inter-connected through belonging to this sacred unity. An example is St Paul's notion in Colossians 1 that all things cohere in Christ. It seems reasonable to say that modern biology is belatedly lending some modest support to the mystical experience of the unity of all things. Admittedly, as biology sees it, everything is not equally inter-connected with everything else; there seems to be a tendency for us to be more affected by "proximal" (nearby) influences than by "distal" (remote) ones. The mystical vision of unity thus goes further than biology, and has a more absolute vision of inter-connectedness.

It is notable that there is also interdependence in St. Paul's vision of the church, as set out in 1 Corinthians 12, in which the different members of the church are dependent on one another in the use of their different gifts, as the parts of a body are dependent on each other. People with different gifts in the Christian community, such as prophets and pastors, need each other, he says, as much as different parts of the body such as eye and hand need each other. This is a strong view of the inter-connectedness within the church, and St Paul's view of it is analogous to the inter-connectedness that ecology is discovering in the living world.

The emphasis on context and interdependence in the living world that is emerging from biology also speaks to the challenges currently facing humanity. I suggest that a key challenge facing humanity is to discover an interdependence in human social life that is similar to that found in the living world. Interdependence in human life needs to steer a path between excessive autonomy and individualism on the one hand, and excessive control and subservience on the other (Verney, 1976). Indeed, reaching such interdependence may be central to God's purpose for humanity, and for creation more broadly. Humans often seem to be locked in a battle between autonomy and authority, a battle that gives rise to so many human problems and which can only be resolved by a recognition of interdependence. I suggest that biology is increasingly revealing to us the interdependence that is God's purpose for humanity.

We may see here a new kind of natural theology (or theology of nature) arising from the living world studied by biology. Historically, biology has been especially important in British natural theology, but it is biological natural theology that has been especially undermined by Darwinism (Olding, 1991). Natural theology now focuses largely on the fine-tuning of the universe. It is arguable that natural theology has become too focused, especially since Paley, on providing arguments for the existence of God. I suggest that, for those already working with a theistic metaphysics, the systemic inter-connectedness of the living world may provide a revelation of an important aspect of the nature of the Godhead, and a revelation of God's purposes for humanity. It can be read as a God-given challenge to achieve in human life the kind of interdependence revealed in the biological world.

Conclusion

The concept of laws of nature is questionable on general grounds. It implies an exaggerated view of the lawfulness and predictability of nature, and reifies the laws to which nature is supposed to be conforming. It also sits particularly uneasily with biology, especially in the light of the recent emphasis of the biological sciences on complex, systemic, contextual explanations. This chapter has explored some of the theological issues raised by that new biology. First, the new biology is inconsistent with complete reduction to lower-level processes, and resolves what has been a troublesome issue on the interface of science and religion. Second, it suggests a more subtle and holistic view of God's action in the world, rather than seeing it in terms of God acting at a discrete point to overturn the laws of nature. Third, I suggest that the new biology reveals an interdependence in creation that reflects the *perichoresis* within the Godhead, and may be God's purpose for creation.

References

Brown, W., Murphy, N. & Malony, H.N. (eds.) (1998) *Whatever Happened to the Soul?* Minneapolis, MN: Fortress Press

Clayton, P. (2006) *Mind and Emergence: From Quantum to Consciousness*. Oxford & New York: Oxford University Press

Crick, F. (1994) *The Astonishing Hypothesis: The Scientific Search for the Soul*. London: Simon & Shuster

Dawkins, R. (1976; 2nd edn 1989) *The Selfish Gene.* Oxford & New York: Oxford University Press

Dennett, D. (1995) *Darwin's Dangerous Idea: Evolution and the Meanings of Life.* New York: Simon & Shuster

Denton, M.J. (1998) *Nature's Destiny: How the Laws of Biology Reveal Purpose in the Universe.* New York: The Free Press

Francis, R.C. (2012) *Epigenetics: How Environment Shapes Our Genes.* London: W H Norton

Goodwin, B. (1994) *How the Leopard Changed its Spots.* Princeton: Princeton University Press

Grey, M. (2003) *Sacred Longings: Ecofeminist Theology and Globalisation.* London: SCM Press

Knox. C. (1993) *Changing Christian Paradigms*. Leiden: Brill

Morris, S.C. (ed.) (2008) *The Deep Structure of Biology.* West Conshohocken, PA: Templeton Press

Noble, D. (2006) *The Music of Life: Biology beyond Genes*. Oxford & New York: Oxford University Press
Olding, A. (1991) *Modern Biology and Natural Theology*. London: Routledge
Peacocke, A. (2001) *Paths From Science Towards God: The End of All Our Exploring*. London: Oneworld
Polkinghorne, J. (1998) *Belief in God in an Age of Science*. New Haven: Yale University Press
Prigogine, I. & Stengers, I. (1984) *Order Out of Chaos*. New York: Bantam
Ruse, M. 2010. *Science and Spirituality: Making Room for Faith in an Age of Science*. Cambridge & New York: Cambridge University Press
Soulen, R.K. & Woodhead. L. (eds.) (2006) *God and Human Dignity*. Grand Rapids, IL: Eerdmans
Teske, J.A. (2013) "From embodied to extended cognition". *Zygon*, 48 (3), 759-787
Verney, S. (1976) *Into the New Age*. London: Fontana/Collins
Waddington C.H. (1942) "The epigenotype," *Endeavour* 1, 18–20. (Reprinted in International Journal of Epidemiology 41, 10–13, 2012).
Watts, F. (2011) *Spiritual Healing: Scientific and Religious Perspectives*. Cambridge: Cambridge University Press.
Woollett, K. & Maguire, E. A. (2011). Acquiring "the knowledge" of London's layout drives structural brain changes. *Current Biology*, 21 (24), 2109-2114.

CHAPTER SEVEN

LAWS OF MATHEMATICS: NATURE AND ORIGINS

GAVIN HITCHCOCK

Ask a mathematician for an example of a "law", and you may be offered the *commutative law* or perhaps the *distributive law*. It was Servois who first introduced the words "commutative" and "distributive" into mathematics in connection with functions, in 1814. The more subtle *associative law* was recognised and named only in the 1840s. Mathematics students today meet these words in connection with the arithmetic operations of addition and multiplication: each operation is commutative and associative, and multiplication is distributive over addition.

Laws of Arithmetic

Associative laws:	$(2 + 3) + 5 = 2 + (3 + 5)$
	$(2 \cdot 3) \cdot 5 = 2 \cdot (3 \cdot 5)$
Commutative laws:	$2 + 7 = 7 + 2$
	$2 \cdot 7 = 7 \cdot 2$
Distributive law:	$2 \cdot (3 + 5) = 2 \cdot 3 + 2 \cdot 5$
Law of moduli:	$\vert(-2)(-5)\vert = \vert-2\vert\,\vert-5\vert$

Later, students are introduced to the laws in general symbolic form – the only form in which they can be (and were) recognised and propounded as *laws*, not just procedures.

Laws of "Symbolic Algebra"

Associative laws:	$(a + b) + c = a + (b + c)$
	$(ab)c = a(bc)$
Commutative laws:	$a + b = b + a$
	$ab = ba$
Distributive law:	$a(b + c) = ab + ac$
Law of moduli:	$\|ab\| = \|a\|\,\|b\|$

The first person to propose the naming and study of such laws in general symbolic terms was George Peacock, in 1830. A respected mathematician and member of the Cambridge and Anglican establishment, his moment came to send intellectual shock waves through the world of nineteenth century British mathematics, and initiate a far-reaching revolution in algebra. He had agonised for some years over the embarrassing lack of foundations of algebra, in which the negative and imaginary numbers were often branded as outlaws. Then, in 1830, with a strange combination of solemnity and anxiety, he produced his *Treatise on Algebra,* an attempt to give to the much-maligned subject the dignity and stature of a true deductive science, like geometry.

SCENE: High Table at Trinity College, Cambridge, 1831

Dramatis personae

- Christopher Wordsworth (1774-1846), Master of Trinity College, and Cambridge establishment figure. Aged 57, he is the brother of the poet William Wordsworth.
- George Peacock (1791-1858), Fellow of Trinity College. Aged 40, he is author of *A Treatise on Algebra.*
- William Whewell (1794-1866), Peacock's contemporary and colleague in Trinity College. Aged 37, he will later author *The Philosophy of the Inductive Sciences Founded Upon their History,* and coin words such as "scientist", "physicist".
- Osborne Reynolds, young mathematician. He will be Twelfth Wrangler in 1837, and write an anonymous pamphlet criticising Peacock's *Algebra.* After holding a fellowship at Queen's College, Cambridge, he will pursue careers as rector and schoolmaster. (His son of the same name would be a good mathematician and famous engineer.)

Note: The words in this dialogue are closely based on primary sources, slightly edited for the stage, except for those of Wordsworth, who is my own representation of Cambridge establishment views at the time. The sources are (Peacock, 1830), (Whewell, 1835), and, for Reynolds' words, (Pycior, 1882).

WORDSWORTH: Well, Peacock, what is this nonsense I hear about your book on Algebra? – Seems to be stirring up a hornet's nest in Cambridge!

PEACOCK: I fear that is true, Master – allow me to explain myself. I consider that Algebra, in its most general form – I call it "Symbolical Algebra" as opposed to "Arithmetical Algebra" – is the science which treats of the combinations of *arbitrary signs and symbols*, by means of *defined though arbitrary laws*.

WORDSWORTH: Oh, dear me! Quite intolerable! Such talk of assuming any laws we like is an abomination. How can *arbitrarily* imposed rules on *arbitrary* symbols have any meaning at all?

PEACOCK: It's true that the symbols are *perfectly* general in their representation, and the operations are *perfectly* unlimited. The laws of these operations are *altogether independent* of the specific meanings of the symbols themselves.

WORDSWORTH: But this is an *outrageous* position for a mathematician! How can you assume laws when you have no idea what you are talking about?

PEACOCK: We may assume *any* laws for the combination and incorporation of such symbols, so long as our assumptions are *independent* and therefore not inconsistent with each other.

WORDSWORTH: And what do the younger mathematicians think of this monstrous concoction pretending to be Algebra? Eh, young Reynolds? – Speak up!

REYNOLDS: Well, Dr Peacock certainly has his disciples … but, having struggled with the book, I confess to grave misgivings. Surely, to be a true science, Algebra must have content or meaning? Interpretation of the signs must *precede* rather than follow their manipulation.

[turns to Peacock] To be frank, sir, I believe that your symbolical algebra is devoid of meaning. Therefore it is unworthy of the name of science.

PEACOCK: In symbolical algebra, the rules determine the meaning of the operations; they are arbitrarily imposed upon a science of symbols and their combinations.

REYNOLDS: But you appear to contradict yourself, sir. On the one hand you claim that the symbols are intended to represent nothing but themselves, and are symbols *only*, in meaning as in appearance. On the other hand, in your Algebra you say that the symbols denote indifferently *every species* of quantity, abstract and concrete.

PEACOCK: Indeed, the admission of this principle of the independent existence of the operations makes it necessary to consider symbols, not merely as the general representatives of numbers, but of *every* species of quantity. One of the most important consequences of this view is the *complete separation* which it effects, of the laws for the combination of symbols from the principles of their interpretation.

REYNOLDS *[gesticulating]*: A symbol is *nothing*, until some representation is given to it. A symbol as such is not susceptible of *any* operation. What is the meaning of adding **a** to **b**, if **a** and **b** be *nothing* more than their forms designate? To speak of a mathematical operation on a symbol as such *only*, does *violence* to our ideas of things!

WHEWELL *[laughing]:* Bravo, young Reynolds! Symbolical Algebra! Mumbo-jumbo! Peacock, you would confront us with such a spectacle as pagan symbolistic rituals, signifying nothing.

WORDSWORTH: This so-called Symbolical Algebra is not mathematics. My dear Peacock, *things* – most especially *mathematical* things – are what they *are*; they behave the way they behave, and our task is to find out *how* they behave and treat them appropriately, with all the respect due to the concepts of an *exact* science!

PEACOCK: Not so exact, I must point out! It is on account of the unhappy state of the doctrine of negative and imaginary quantities, which have perplexed analysts for so long, that I have felt myself compelled to depart so very widely from the form in which the principles of Algebra have been commonly exhibited. The object I propose to effect is undoubtedly one of great importance, and of no small difficulty, and will cause much dispute and controversy …

WHEWELL: Peacock, when we were younger, and fellow revolutionaries in the analytical cause, I admired your courage in taking on the die-hards of Cambridge. But now, with respect, I cannot but *deplore* mathematics taught in such a manner that its foundations appear to be laid in *arbitrary* definitions without any corresponding act of the mind! Or mathematics held forth as if its highest perfection were to reduce our knowledge to extremely general propositions and processes. For such mathematics unfits the mind for dealing with other kinds of truth.

PEACOCK: Whewell, my dear fellow, may I make bold to suggest that Symbolical Algebra is a step towards the next stage of our analytical revolution? I am very sensible of the great responsibility which I incur by the proposal of so many innovations. But my aim is to remove any difficulties or imperfections from the elements of this beautiful and most comprehensive science, and I can only hope that posterity will vindicate my much-criticised presumption!

WHEWELL *[genially]:* Ah, well, we shall see, we shall see! Let us return to more amiable conversation, and drink to the good sense of the next generation of Trinity mathematicians!

[All raise their glasses ...]

From frank puzzlement to future law making & breaking

Peacock's Algebra precipitated a profound debate about the very nature of mathematics. His students at Trinity College included some, like Duncan Gregory and Robert Ellis, who embraced his ideas and formed a school of Symbolical Algebra that nurtured Arthur Cayley and George Boole. But two of the brightest young mathematicians of the day were at first very puzzled indeed. Augustus De Morgan, another of Peacock's students, graduate of Trinity College and first Professor of Mathematics at the fledgling University College of London, delayed reviewing his mentor's book until he felt he could make some sense of it:

> [I should explain why this book] has not received a notice before now, seeing that it was published five years ago. The reason is the very great difficulty of forming fixed opinions upon views so new and so extensive. At first sight it appeared to us something like symbols bewitched, and

> running about the world in search of a meaning. [...] However, the work of Mr Peacock, though difficult, is logical. (De Morgan, 1835)

And William Rowan Hamilton (1805-1865), an Irish child prodigy who won all the prizes and honours that the other Trinity College (in Dublin) had to bestow and was appointed Astronomer Royal of Ireland when only 22, recalls in 1846 his early revulsion:

> [It seemed to me then that Mr Peacock] designed to reduce Algebra to a mere system of symbols *and nothing more*; an affair of pothooks and hangers, of black strokes upon white paper, to be made according to a fixed but arbitrary set of rules. I refused to give the high name of Science to the results of such a system. (Graves, 1882-89)

But Peacock had opened the road to abstract algebra, by creating a transitional framework of discourse – an algebraic language in which, for the first time, laws could be articulated and compared. Each of these two men was deeply influenced, and about a decade later each made algebraic history. De Morgan (1842) tossed off an almost complete list of what we now regard as the field axioms, capturing the behaviour of real numbers and complex numbers:

> As far as I can see (and I believe no writer has professed to throw together in one place everything that is essential to the algebraic process), the laws of operation are as follows [...] I believe the preceding rules to be neither insufficient nor redundant ...

Hamilton shocked the scientific community by announcing in 1843 his discovery of *Quaternions,* the first truly novel system to be created and recognised as an algebra. Peacock and his school of symbolical algebraists had asserted the freedom to decree arbitrary laws, but none went so far as to actually cut the umbilical cord and free the baby from the mother system, Arithmetical Algebra. Hamilton himself was far from practising the freedom Peacock had preached, but the cognitive pressure of his sustained intellectual mission built up until a moment, now legendary, when walking with his wife along the Royal Canal into Dublin from the Observatory:

> An *electric* circuit seemed to *close*; and a spark flashed forth ... Nor could I resist the impulse – unphilosophical as it may have been – to cut with a knife on the stone of Brougham Bridge, as we passed it, the fundamental formula with the symbols i, j, k ... which contains the *Solution* of the

Problem, but of course, as an inscription, it has long since mouldered away. (Hamilton, 1865) /'

Hamilton's engraving:

$$i^2 = j^2 = k^2 = ijk = -1$$

Definition of a Quaternion:

$$Q = a + ib + jc + kd,$$
$$\text{where } ij = k, \quad ji = -k, \quad etc.$$

The Law of Moduli:

$$\text{If } Q_1 = a + ib + jc + kd$$
$$\text{and } Q_2 = \alpha + i\beta + j\gamma + k\delta,$$
$$\text{then } |Q_1 . Q_2| = |Q_1|.|Q_2|$$

Hamilton's mental journey had involved a complex bargaining with the making and breaking of laws (Hamilton, 1853), and he was finally forced to imagine into being three new "imaginaries": *i, j, k,* forming a four-dimensional system, and breaking the commutative law, whilst behaving impeccably in other important respects such as the law of moduli. His intellectual odyssey was framed in the language of Peacock the prophet, who provoked debate about laws. As the Apostle Paul has it: "Without law there is no concept of sin": no moral compass, and similarly no mathematical compass. Hamilton's Quaternion system was finally recognised as a valid and useful algebra, and played an important role as precursor of vector calculus.

New algebras

Amidst much initial consternation and scepticism, Hamilton's discovery opened the floodgates for a stream of strange but beautiful new algebras, at first called "number-systems'. Though still concrete (not yet abstract), these discoveries, embraced by British mathematicians, prepared the way for the twentieth century apotheosis of algebra into abstract algebra. Matrix algebra and group algebras were introduced in the 1850s by Arthur Cayley. What is now called "Boolean Algebra" was introduced in 1854 by George Boole as *The Laws of Thought*, an event that Bertrand Russell claimed signalled the greatest discovery of the nineteenth century – pure mathematics. Here are some of the new algebraic structures, many recognisable as important systems today, others less known, but all contributing to a fundamentally different view of mathematics:

1843: HAMILTON, *quaternions*
1844: JOHN T GRAVES, *octaves, octonions,* or *Cayley numbers*
1844-5: DE MORGAN & JOHN GRAVES, *triple algebras*
1846: CHARLES GRAVES, *a triple algebra*
1847: DE MORGAN & GEORGE BOOLE, *formal logics*
1848: THOMAS KIRKMAN, *pluquaternions*
1848-9: JAMES COCKLE, *tessarines, coquaternions*
1853: HAMILTON, *biquaternions, hypernumbers*
1854: BOOLE, *laws of thought*
1854-8: ARTHUR CAYLEY, *group algebras, matrices*
1873: WILLIAM CLIFFORD, *Clifford algebras*
1882: JOSEPH SYLVESTER, *nonions*

Meanwhile, Continental mathematicians had been exploring and proving theorems in concrete systems they did not yet perceive as algebras, noting "affinities" they were not yet ready to pursue, and glimpsing regularities they were not yet in a position to articulate, compare and contrast as laws. Gauss glimpsed what we call "groups" in his arithmetic of residue classes, and classes of quadratic forms. Abel and Galois glimpsed groups and finite "fields"; Kronecker glimpsed "Abelian groups"; Dedekind glimpsed fields. But before these could be perceived as instances of new species of algebras, and incorporated in abstract mathematical theories of great beauty and generality, something, still to be supplied by the British, was missing.

Some of these new systems were rejected out of hand as outlaws. When laws that had been previously assumed universal were broken, the response could sound like moral outrage:

> It is greatly to be lamented that this virtue of the real integers, that they can be decomposed into unique prime factors, does not belong to the complex integers. For, were this the case, the entire theory, which is still labouring under many difficulties, could be easily resolved and brought to a conclusion. (Kummer, 1847)

Two leading French mathematicians later expressed disgust at the new analysis, generating counter-intuitive, lawless monsters:

> I turn away with fright and horror from this lamentable evil of functions which do not have derivatives! (Hermite, 1893).

> Logic sometimes makes monsters. For half a century we have seen a mass of bizarre functions which appear to be forced to resemble as little as possible honest functions which serve some purpose. More of continuity,

> or less of continuity, more derivatives, and so forth ... In former times when one invented a new function it was for a practical purpose; today one invents them purposely to show up defects in the reasoning of our fathers, and one will deduce from them only that (Poincaré, 1897).

The journey to abstraction

While, in Britain, algebra was rapidly becoming *algebras*, on the Continent geometry was fitfully becoming *geometries* – not in a community debate about laws, but in a series of isolated excursions by individuals reasoning from newly adopted, no longer "self-evident", postulates. This is a very different story from that of algebra, but the gradual emergence of abstract algebra and non-Euclidean geometries was characterised by a common theme – the delighted apprehension of ravishingly beautiful though weird new worlds, and the beginnings of their classification by enunciating laws, following Peacock's "laws of combination of symbols". As early as the 1870s, this transcendence was underway, demanding a completely new level of perception and nomenclature of structures. The abstraction process was greatly aided by the set theory of the German mathematician, Georg Cantor (1845-1918), and inspired by his daring presumption in setting out to codify the laws of the infinite.

In the twentieth century, the new *structural algebra* became *abstract algebra,* geometry embraced *Hilbert space, Euclidean spaces,* etc., and calculus became *analysis*, and *abstract analysis.* Disapproval of rampant lawlessness gave way to wonder at the god-like capacity and audacity of mathematicians to decree worlds into being. And then – glorious surprise! – it turned out that the weirdest of these provided ideal model worlds for physics. Non-commutative algebra, non-Euclidean geometry and abstract Hilbert space provided the natural frameworks for relativity and quantum theory.

Ever since, the mathematical community has encouraged a balance between freedom and groundedness, recognising that the best and the most fruitful mathematics arises from a disciplined imagination – an exquisite poise between mathematical freedom and humility before the way the world is, guided by a well-tutored sense of "rightness" and aesthetic beauty. Hamilton's creatures got away with breaking the commutative law because they were so beautifully behaved in other ways. The mathematician's aesthetic judgement is crucial in choosing axioms, seeking and naming "lawfulness", and conjecturing theorems, just as the scientist's intuition and instincts are crucial in choosing which elements of the natural world should be retained in the ideal representation of the problem, and which should be discarded as unimportant details.

Engraved on Georg Cantor's memorial are these words (attributed to him), encapsulating the dawn of a new approach to mathematics:

> The essence of mathematics is its freedom.

(This and the two "quotes" below are part of mathematical folklore.) Henri Poincaré, around 1910, vividly expressed a natural worry about the apparent licentiousness and uselessness of such abstract excursions:

> Later generations will regard Cantor's set theory as a disease from which one has recovered.

But the German mathematician David Hilbert (1862-1943) thundered prophetically in response, affirming (what the mathematical community has defended ever since) that there is great beauty, and intrinsic worth and goodness, in such mathematics:

> No one shall expel us from the paradise that Georg Cantor has opened up for us.

Conclusion

This has been, briefly, the story of the emergence (with algebra leading the way) of a treasury of abstract forms and structures that are studied and enjoyed for their own sake, while being available for the modelling (often unexpected) of physical reality. In science, mathematical model worlds are the means to abstract and express patterns and regularities observed in nature; in nineteenth century structural algebra, *mathematical laws* became the means to describe patterns and regularities observed *across mathematical structures.* The British recognition and creation of new algebras, and the characterisation of species by distinctive laws of composition, made possible a comparative anatomy and taxonomy of species. Mathematical laws helped to make precise the degree of conformity between structures, through what we now call *isomorphism.* Thus were named the elusive affinities that could then be abstracted in precise axioms defining abstract groups, rings, fields, etc.

George Peacock, in his pioneering exposition of a provisional form of thought and symbolisation whose fruits we take entirely for granted today, encountered disbelief and outrage. This moment in the history of mathematical thought, portrayed here in dialogue form, signals the emergence of a framework of discourse completely novel and greatly influential. The difficult task of prizing apart the ideas of algebraic form and content was a prerequisite, both for the speculative construction, analysis and comparison of new mathematical systems for their own sake, and also for the apt expression and modelling of the laws of nature. The wonder is that two such different enterprises, often pursued by different sets of people, should be so fundamentally intertwined. This "unreasonable effectiveness" of mathematical thought (as memorably affirmed by Eugene Wigner) gives cause for gratitude – and, for theists, praise, acknowledging the comprehensibility of the world to human minds as an aspect of the *Imago Dei.*

It is common to talk of the beauty of mathematics, reflected in the elegance and economy of physical laws. For a mathematician, there are strong affinities with the psalmist's rapture at the beauty of the moral law: "sweeter than honey ... delighting the soul". The affinities go deeper than simply beauty. The idea of "goodness" expresses well the apprehensions of mathematicians and scientists, and firmly grounds their passions. To use the resonances and metaphors of the Book of Genesis, the human pursuit of mathematics, too, is "Very Good'. Mathematicians create and explore many lovely systems that may not yet have found concrete "animals in the garden" to name, but are valued by other criteria that have evolved over the last century (Hitchcock, 2009). The dance of the universe and our capacity to comprehend its patterns and laws, together with the invitation to create mathematical poetry in the process, may be perceived as a gift and expression of divine love.

Acknowledgements

The dialogue in this paper was excellently performed, impromptu, at the Leeds meeting by four conference participants, and the short individual quotes were also enunciated with appropriate accents and theatrical gusto by others. I wish to express here my gratitude to them all.

References

De Morgan, A. (1835). "Review of George Peacock, A Treatise of Algebra [1830]," *Quart. J. Educ.* 9, 91-110 & 293-311

—. (1842). "On the Foundations of Algebra, No. II," *Trans. Camb. Phil. Soc.* 7, 287-300

Graves, R. P. ([1882-9] 1975). *Life of Sir William Rowan Hamilton* (3 vols). Dublin: Hodges, Figgis & Co., reprinted New York: Arno. Available online at: http://openlibrary.org/books/OL6952298M

Hamilton, W. R. (1853). *Lectures on Quaternions*. Dublin: *Publisher?*

—. (1865). Letter to his son Archibald, reproduced in Graves (1882-89), vol. 2, 434-5. Italics represent Hamilton's original underlining.

Hitchcock, A. G. (2009). "Cosmic conversation: the evolving dialogue in mathematics between mind and reality", in N. Spurway (ed.) *Theology, Evolution, and the Mind*. Newcastle: Cambridge Scholars Publishing, 178-188

Hermite, C. (1893). Letter to Stieltjes, quoted in Kline (1990), 973.

Katz, V. J. (2009). *A History of Mathematics: An Introduction* (edn 3). New York: Pearson Addison-Wesley

Kline, M., 1990. *Mathematical Thought from Ancient to Modern Times*. Oxford: Oxford University Press

Kummer, E. E. (1847). "De numeris complexis, qui radicibus unitatis et numeris integris realibus constant," *J. mathématique pures &appliqués* (= Liouville's journal), 12, 185-212, quoted in Katz (2009), 709

Peacock, G. ([1830] 1940). *A Treatise on Algebra*. Cambridge: J. Deighton. Reprint edition: New York: Scripta Mathematica.

Poincaré, H. (1899). *L'Enseinement mathématique* 11, 157-62, quoted in Kline (1990) 973.

Pycior, H. M. (1981). "George Peacock and the British origins of symbolic algebra." *Historia Math*. 8, 23-45.

—. (1982). "Early criticism of the symbolical approach to algebra." *Historia Math*. 9, 392-412. See particularly her extracts and paraphrases (402-7) of Reynolds' officially anonymous pamphlet: *Strictures on Certain Parts of Peacock's Algebra*, Cambridge (1837).

Whewell, William, 1835. *Thoughts on the Study of Mathematics as a part of a Liberal Education*. Cambridge: Pitt Press (1836).

CHAPTER EIGHT

THE LINK BETWEEN THE CONCEPTS OF LAW OF NATURE AND CONTINUOUS CREATION

FABIEN REVOL

The Link between the Bible and the "Law of Nature" Concept

The idea of a law of nature arises with modern physics from the XVIth century. The notion of law applied to nature appears in a context wherein religious faith gives the picture of a God imposing or applying his law onto the elements of creation understood as a machine. This notion is thus dependent on that of divine law. This is the human experience of God the legislator for social life which, by analogy, leads us to think of that parallel with nature. Johannes Kepler seems to be the first scientist to use the term in the modern perspective, particularly in optics. Neither Galileo Galilei nor Blaise Pascal use it. However, it is René Descartes who is recognized as the creator of the concept of a law of nature: he "combined quantitative rules used by craftsmen and engineers of his time with the biblical concept of God the Legislator" (Jaeger, 1999). In *Le Monde*, Descartes exemplifies the use of the laws for the organization of the cosmos by God (Descartes, [ca 1633] 1998). And in his letter of the 15th April 1630 to Mersenne he clearly writes: "It is God who has established these laws in nature, like a king establishing laws in his kingdom" (Descartes, 1997).

Newton consolidated this concept in his *Principia Mathematica* (Newton, [1713] 1999). From him the use of the concept of laws of nature was for a while generalized to all scholars. It is thus important to note that the world understood as creation is the substrate from which could grow the concept of a law of nature. This first approach allows the understanding that the themes of creation and of laws of nature are intimately connected. It is thus to Descartes that we owe the conceptualization of the essence of those laws. To be more precise, there is in Descartes a dependence on this concept alongside that of continuous creation. One wonders if this connection is still accurate: what concept of a law of nature could fit with what concept of continuous creation?

For Descartes: an opportunity to elaborate the concept of "law of nature"

With Descartes, the distinction between creation and conservation fades, giving way to the concept of "continuous creation" with the idea that keeping things in being requires the same kind of action as that which gives them being initially. In the third metaphysical meditation Descartes argues that the distinction between creation and conservation is purely conceptual. "So that natural light makes us see clearly that conservation and creation differ only in regard to the way we think and not in effect" (Descartes [1641] 2013). The problem is that if God created without conserving, creation would seem futile. However, Paul Clavier has commented that for God, according to Descartes, being cause of the object *x* at time *t* is both creating and conserving. "If the effect of the act of creating is durable, so the distinction between creating and conserving shrinks: the creative cause exercises the power to conserve in existence (or in the existing condition) the creature" (Clavier, 2011).

Here one can draw a parallel with Thomas Aquinas in the sense that, for Thomas, creation is more a matter of dependency than of beginning. The creative relationship founds the created object in the present and in the relationship with his creative cause. It is the same for Descartes. For him, continuous creation is a divine concourse "that supports the creatures in existence at any time" (Valle, 1971). But we will see that the method of operation of such a creation is not at all the same for Descartes as for Thomas.

Descartes borrows from Judeo-Christianity the idea of a transcendent creator God that he adapts to the system of his physics. It then goes into the break with a biblical reflection on the origins (Valle, 1971). The justification of creation-conservation is to be found elsewhere: in a reflection on the existence of God as the foundation of the existence of the thinking subject. In the third meditation, where we find the operation of continuous creation, Descartes uses the ontological proof of God's existence: God exists because I have been given the idea of God and his perfections: however a being so perfect could not *not* have this first perfection, which is existence. He cannot mislead me because if he did, he would not be perfect and would therefore not be God. God exists because if he did not exist I could not have the idea as I have it, with his perfections. But what implications would it have for my life if God did not exist? Descartes replied that I would be God. But as I do not have the divine perfections, and I am not able to assign them to myself, I am not God, so it is necessary that God drew me out of nothing and made me exist (Descartes [1641] 2013).

It is already clear that continuous creation has, for Descartes, a utility function as evidence of the existence of God. It is also a step in Descartes's argument to establish the thinking subject. It also has a different function in relation to his system of knowledge of nature.

For him, continuous creation is a divine concourse "that supports creatures in existence at any time" (Valle, 1971). The theme of continuous creation is thus introduced in the third meditation: the existence of my life is infinitely divisible into parts that are not dependent on each other; being at one time does not involve being at another. So at each of these moments God makes me exist and "conserves me." (Descartes [1641] 2013) At each of these moments my existence is independent of that at the previous moment. Every time I am re-created anew by the same creative power. In summary, for Descartes continuous creation implies a creative act every moment *t* of passing time; this creative act recreates everything again. The reason for this re-creation at every moment is found in a reflection on causality and substance in opposition to the viewpoint of scholasticism. Since an existing entity no longer has a formal substance, as cause of its being and its becoming, Descartes offsets this by the direct, immediate and ongoing support of God through constant creative reenactment. Nature is therefore totally inert and incapable of producing anything by itself that is not given by God. So much so that, in the third meditation, Descartes thinks that the state of creation at time *t* does not depend on the previous time and is totally independent of the creative action of this very instant. Consequently, matter is only inert, extended in space, perfectly knowable by geometric relationships, that is to say by science. The world being emptied of its substance and of its natural causality, there are only regular movements, which are a testimony that God does not change his mind and the laws of motion are permanent. The divine immutability guarantees the laws of nature. Mathematics can also exhaust the reality that can be understood in terms of laws.

Paul Clavier confirms this view, highlighting the link between ordering and creation in Descartes thought, a link that illustrates the idea of laws of nature. In the Sixth Meditation, Descartes says that God establishes the creation in a coordinated manner. The imposition of laws upon nature contributes to the idea that God brings its concourse to nature in a gradual uplifting from chaos. "Nature exists only by virtue of the conservative action and acts only because God makes it act" through laws imposed or printed on it (Clavier, 2011). A law of nature is somehow the immanent activity of God organizing creation.

Blaise Pascal's criticism of Descartes for claiming that God required just one "flick" to create all the matter, energy and movement in the Universe (Pascal [1669] 2007) seems to be unjustified. Descartes' God is not a God who leaves the creation on its own once it has been created and received the package of laws necessary for its operation. God creates everything by giving it a force, that it is a resting force or one of movement. God is the source of all movement and of everything else in nature, it is guaranteed by his ordinary concourse that keeps each of these things in movement or at rest. The laws of motion do not change; this guarantees stability and motion as well as our possible knowledge of them. This state is analogically guaranteed by the fact that God does not change.

In the *Principles of Philosophy*, Descartes also identifies the laws of nature with the secondary causes of movements (Descartes [ca 1633] 1988). The conclusion of the *Principles* is clear, there is balance between the ability to know something as created by God, and the permanence of the world, as it is preserved by God. This conservation is a guarantee of a world knowable by human intelligence because God does not fail us (Descartes [ca 1633] 1988).

Changes in the sciences since Descartes have led to a still unresolved epistemological debate on the topic of laws which is reflected in much of the rest of this book. However it might also be interesting to speculate on the impact of recent developments in scientific paradigms on the theological presupposition of continuous creation.

A new relationship between natural law and continuous creation

For Lydia Jaeger (2008), Einstein's relativity and quantum physics restore chance and indeterminacy to physics. However, they do not change the concept of the laws of nature. But we can wonder about the nature of the uncertainty of which this physics speaks. If this uncertainty is only epistemological, there is perhaps no reason to change anything about the concept of a law of nature. If it is ontological, it is likely that the concept of a law of nature is thereby changed. To pursue this point I will follow the reflections of the philosopher Karl Popper on the meaning of probability in statistics.

Popper points out that the statistical distribution or law of a series of events is defined *a posteriori* from a large number of observations. He takes the example of a dice to say that, if the observed frequency of a number in a series of throws is markedly different from 1/6, one concludes that the dice is loaded: one of the sides has been weighted so that the opposite side is as often as possible the upper one when the dice is thrown. The possible results of such a throw are not equally probable, one outcome is loaded. In other words the loaded dice has a propensity to generate the expected event with a greater-than-chance probability. This tendency is said to be stable if, in multiple series of throws, the same average value is obtained as long as the environment where the measurement takes place does not change. That is to say, there must be a set of invariants that supports that stability: "the field of propensity that influences each singular throw" (Popper, 1990). Popper then extends this thinking to all natural events. The concept of dependency of events would seem to indicate that, in nature, some possibilities are so balanced that they are no longer mere possibilities but expectations. In the extreme case, when a cause determines its effect, Popper sees there a particular case of a propensity whose value is equal to one. That is to say that the probability for that effect in the context of the exercise of the cause is equal to one.

Taking a realistic view, science is, according to Aristotle the knowledge of the causes that produce their effects; a law of nature is that which allows us to formulate such knowledge. For Aristotle it was the knowledge of a necessary connection between cause and effect: a deterministic relationship. Popper's approach instead represents the relationship in probabilistic terms. A necessary relationship, constituting a law of nature, corresponds to the exercise of a cause with probability of effect equal to one. This suggests to Popper that there exist in nature causes which do not produce their effects with such a high probability, and the more the probability moves away from 1 towards 0.5 – the probability of greatest uncertainty – the more there are exceptions to the law that seeks to establish a precise relationship between the effect and the cause. The shift in perspective from a nature operating under deterministic and mechanistic laws to a nature operating under probabilistic, statistical laws can introduce an ontological uncertainty into the natural phenomena studied by science.

However it must be emphasized that Popperian propensity is defined in a context within which there is stability. This means that the propensity is not an inherent property of an object alone, but of a situation: it depends on a network of interdependence. Popper has developed a fairly strong conclusion: "determinism is simply not true" (Popper, 1990). Propensities open doors to the register of indeterminacy in science and nature. Past situations do not *determine* future situations; rather, propensities *influence* future situations. In this perspective, the future is open and the unknown of the becoming is not related to a lack of knowledge of the causes, but to the probabilistic nature of the actualization of a propensity, an open actualization submitted to the modification of the propensity itself. This means that what we call “invariants” or “regularities of nature” are not absolute. What appears to be invariant is that way because of the extreme stability of a field of propensities, that is to say, with regard to the laws of physics, that the scope of the propensities which determine them has been stable since the Big Bang and will continue to be stable during the lifetime of our universe. On the other hand, other fields of propensity may change the conditions for expression of natural phenomena and thus allow causes to express themselves more or less deterministically, such that new causalities are expressed. So, with Popper, we can say, "This universe of propensity of ours is inherently creative” (Popper, 1990).

In the same vein, Arthur Peacocke has absorbed Popper’s insights into his theology of continuous creation (Peacocke, 1979, 1993, 2004). For Peacocke, the notion of propensity expresses trends that some potentials from among many can be actualized because of the environmental pressures on a system. The propensity describes a situation wherein the context will bring about the result that a random phenomenon becomes more than likely. Natural selection seems to belong to this order in the living world: the environment exerting selective pressure on organisms promotes the emergence of stronger or more complex characteristics that allow better adaptation.

This leads us to think of a concept of continuous creation that is not merely conservation in being, ensuring the stability of the laws of nature. Instead, we have considered the relative stability of these laws in the field of propensities. This leads us to consider a divine creation that takes place within natural creativity.

For today, according to the shift of paradigms involving the concept of a law of nature, the meaning of this concept is taken neither from Scholastic (Suarez ([1597-1636] 2002) nor from Cartesian traditions (Descartes [1641] 2013). The concept of continuous creation is nowadays widely used in reflection on the dialogue between science and religion. It is a modern idea that divine creative power is involved within temporality. In the dialogue between the theories of evolution and theology, this viewpoint sees natural novelty as the fruit of a creative act that involves both creator and creature without any breaking of the natural functioning of evolution, without any external interfering of God’s action within creation as observed by science, but truly as the immanent sustainer of those natural processes. It is a concept well fitted to express the dynamic aspect of nature, read and interpreted through the teachings of the story of life on earth, according to the framework of the theory of evolution. Continuous creation is the collaborative interaction of God, by His Wisdom and Spirit, with creation itself. This collaboration allows the permanent and continuous emergence of novelty in the world, in the form of the procession of creative information, offered by God and discriminated by the creatures. This information comes from the invisible world and is established as the universe of possibilities of creative paths, of which the divine Word is the model and exemplary cause under the creative will of the Father. Those possibilities are not necessities (Revol, 2013a,b).

All creative possibilities are not expressed in creation. Only a few are, according to the creative dialogue that allows or does not allow for a given possibility to be expressed. There is a competition among possibilities for expression in creation. Creatures participate in the process of screening of the possibilities, they are thus co-creators with God (Hefner, 1993). The competition is well expressed in the trial-and-error process of evolution, and in the probabilistic essence of the laws of nature: the propensities and their possible evolution are the result of a way of creation that is open to the rise of true and contingent novelties. As part of an evolutionary interpretation of nature, novelty in life is built as a paradigm of this continuous creation in its contingency, its unpredictability and its ontological consistency.

The process of production of biodiversity implies the concept of constitutive relationship at different levels of intelligibility. On the biological level, ecology teaches us that a being is the emergence of an entity thanks to relationships it has with its natural environment (Ricklefs & Miller, 1999). Natural diversity is the fruit of a complexity of relationships.

On the metaphysical level, the novelty that arises is grounded in the expression of a new substantial form which is created by the organization, arrangement and interaction of previous forms. The natural interactions are the expression of metaphysical interactions. Those interactions are possible because the forms selected for expression have issued from a process in the interactions between God the Creator and the creatures (Revol, 2013b).

Creation is a process understood under the category of relationship, as Thomas Aquinas said ([1274] 2012). It is further understood under the analogy of ecological relationship, according to Jürgen Moltmann (1993), though in different terms and with different backgrounds. But what is interesting here is the idea of relationship that is characteristic of the act of creation, and of the production and creation of novelty. It seems that we have here a new expression of what Saint Bonaventure called the *vestigiae trinitatis*, the traces of God in creation (Bonaventure [1221-1274] 2005). Relationship is an expression of God being as Trinity. For Denis Edwards, continuous creation and, therefore, its expression in the production of novelties such as in the process of biodiversity can be understood as a signature or a footprint of God in creation (Edwards, 1998, 2004).

There is a convergence between the concept of a law of nature deriving from the concept of propensity, and the concept of continuous creation introduced here. If continuous creation is a process wherein contingency is emphasized, through a relational cooperation between God and creation, then, far from being absolutely stable, laws of nature embody in themselves that contingency: they are stable only as long as the network of constitutive relationships works in the framework of a particular propensity.

References

Aquinas, T. ([1274] 2012), *Summa Theologica Part I ("Prima Pars")* Altenmünster: Jazzybee Verlag.

Bonaventure, St. ([1221-1274] 2005), *Works of St. Bonaventure* Vol. IX, *Breviloquium,* D. V. Monti (ed.), Saint Bonaventure: Franciscan Institute Publications.

Clavier, P. (2011), *Ex Nihilo*, vol. 1*, L'introduction en philosophie du concept de Creation.* Coll. Philosophie, Paris: Herman.

Descartes R. (1997), *The Philosophical Writings, Vol. III The Correspondence,* (3rd edn) transl. J. Cottingham, R. Stoothoff, D. Murdoch & A. Kenny. Cambridge: Cambridge University Press, 415.

—. ([ca 1633] 1998), *The World and other Writings*, Cambridge Texts in the History of Philosophy, S. Gaukroger (ed.). Cambridge: Cambridge University Press.

—. ([1644] 1998), *Principles of Philosophy*. Dordrecht: Kluwer.

—. ([1641] 2013), *Meditation on First Philosophy, With Selections from the Objections and Replies*, A Latin-English Edition, ed. & transl. J. Cottingham. Cambridge: Cambridge University Press.

Edwards, D. (1995), "The Discovery of Chaos and the Retrieval of the Trinity", in R. J. Russell, N. Murphy, & A. Peacocke (eds), *Scientific Perspectives on Divine Action,* 2: *Chaos and Complexity*. Vatican, Vatican Observatory Publications & Berkeley, Center for Theology & the Natural Sciences, 157-75.

—. (2004), "A Relational and Evolving Universe unfolding within the Dynamism of the Divine Communion", *in* P. Clayton & A. Peacocke (eds), *In Whom We Live and Move and Have Our Being, Panentheistic Reflections on God's Presence in a Scientific World.* Grand Rapids, MI: Eerdmans.

Frankfürt H. (1987), "Création continuée, inertie ontologique et discontinuité temporelle en Philosophie française", *Revue de métaphysique et de morale* 92, 455-472.

Garber, D. (1987), "How God Causes Motion : Descartes, Divine Sustenance, and Occasionalism. *Journal of Philosophy* 84 (10), 567-581.

Hefner, P. (1993), *The Human Factor: Evolution, Culture, Religion.* Minneapolis, MN: Fortress Press.

Jaeger, L. (1999), *Croire et connaître Einstein, Polanyi et les lois de la nature.* Nogent sur Marne & Cléon d'Andran: Éditions Excelsis.

—. 2008), *Ce que les cieux racontent, la science à la lumière de la Création.* Nogent sur Marne & Cléon d'Andran: Éditions Excelsis – Éditions de l'Institut biblique, 248

Miller, T. D. (2011), "Continuous Creation and Secondary Causation: the Threat of Occasionalism", *Religious Studies* 47, 3-22.

Ibid (2009), "On the Distinction Between Creation and Conservation: a Partial Defence of Continuous Creation", *Religious Studies* 45, 471-485.

Ibid (2007), *Continuous Creation, Persistence, and Secondary Causation: An Essay on the Metaphysics of Theism*, a dissertation submitted to the graduate faculty in partial fulfillment of the requirements for the degree of Doctor of Philosophy, Norman: University of Oklahoma.

Moltmann, J. (1993,) *God in Creation : A New Theology of Creation and the Spirit of God.* Minneapolis: Fortress [first printing 1985].

Nadler, S. (1998), "Louis de La Forge and the Development of Occasionalism : Continuous Creation and the Activity of the Soul". *Journal of the History of Philosophy* 36, 215-231.

Newton, I. ([1713] 1999), *The Principia: Mathematical Principles of Natural Philosophy*, transl. I. B. Cohen & A. Whitman. Berkeley & London: University of California press.

Pascal, B. ([1669] 2007), *Thoughts, Letters and Minor Works.* New York: Cosimo.

Peacocke, A. R. (1979, 2004), *Creation and the World of Science: The Re-Shaping of Belief.* Oxford: Oxford University Press.

—. (2004), *Evolution, the Disguised Friend of Faith, Selected Essays*, Philadelphia & London: Templeton Foundation Press.

—. (1999), "Biology and Theology of Evolution", *Zygon* 34, 695-712.

Peacocke. A. R. (1993), *Theology for a Scientific Age, Being and Becoming – Natural and Divine.* Minneapolis: Fortress.

Popper, K.R. (1990), *A world of Propensities*. Bristol: Thoemmes.
Revol, F. (2013a), "La Création continuée en question", *in* F. Mies (ed.), *Que soit! L'idée de création comme don à la pensée.* Bruxelles: Lessius, 293-303.
—. (2013b), *Le concept de création continuée. Histoire, critique théologique et philosophique, essai de renouvellement dans le dialogue de la théologie avec les sciences de la nature par la médiation de la philosophie,* Thèse pour l'obtention du Doctorat canonique en Théologie, et du Doctorat canonique en Philosophie, Centre Sèvres et Université Catholique de Lyon, 1120.
Ricklefs, R.E. & Miller, G.L. (1999), E*cology,* New York: Freeman.
Sertillanges, A.-D. (1960), *La Philosophie de S. Thomas d'Aquin*. Paris: Aubier, Éditions Montaigne.
Suarez F. ([1597-1636] 2002), *On Creation, Conservation, and Concurrence, Metaphysical Disputations 20,21 & 22*, A. J. Freddoso (ed.). South Bend, Indiana: St. Augustine Press.
Tresmontant C. (1953), *Essai sur la pensée hébraïque*. Paris: Les Éditions du Cerf.
Valle, B. (1971), *La Création continuée chez Descartes*. Mémoire de maîtrise de philosophie, Université de Dijon.

CHAPTER NINE

ON THE COMPATIBILITY OF INTELLIGENT DESIGN AND METHODOLOGICAL NATURALISM

JUUSO LOIKKANEN

Introduction

In the contemporary discussion between science and religion, one of the most controversial phenomena is the theory of intelligent design (ID). Briefly put, the theory holds that "certain features of the universe and of living things are best explained by an intelligent cause, not an undirected process such as natural selection" (Discovery Institute, Center for Science and Culture, 2014). Proponents of ID maintain that some events are too improbable to have come about merely by natural causes, and are thus best explained by an intelligent supernatural designer. Consequently, they claim that since methodological naturalism, which only accepts natural explanations, is incapable of explaining design, it must be abandoned as a basis for science and be replaced by ID.

In this paper, I argue, contrary to ID, that we do not need to set divine design and natural causes against each other. Furthermore, I claim that methodological naturalism still continues to provide the only reliable basis for doing empirical science. However, ID, when understood correctly, is not incompatible with methodological naturalism. Actually, when ID is seen as a theological and philosophical viewpoint rather than as a purely scientific theory, it will contribute more fruitfully to the debate between science and religion.

Intelligent Design

The logical foundations of the theory of intelligent design have been laid by American mathematician, theologian and philosopher, William A. Dembski. He claims that it is possible to differentiate between events that are designed, and events that are caused by something else, by using the so-called criterion of specified complexity. If an event is both highly improbable (complex) and definable as a separate pattern without reference to the actual event (specified), it is – in the sense of Dembski's theory – designed. He argues that this way design can be detected in any kinds of phenomena, including those observed in the biological world. And it is, indeed, in biology that much of the controversy lies.

Dembski (1998; 2002) holds that there are three possible modes of explanation of events occurring in the (what defines an event is, of course, a question of its own): necessity, chance and design. These modes are mutually exclusive and exhaustive, in other words, one and only one of them is the cause of any particular event. Probabilities play a crucial part in determining whether an event is designed or not. If the probability of an event is "high", that is, one or extremely close to one, the event is attributed to necessity. If the probability is "intermediate" – less than one but higher than 10^{-150}, which is Dembski's so-called universal probability bound – the event is caused by chance. If it turns out that the probability is lower than the universal probability bound, the event is complex, and the possibility of design enters the picture.

The crucial step now is to find out whether the event is specified or not. If the event is specified, it features specified complexity and is designed; if the event is not specified, it is caused by chance. Dembski (1998) defines a specified event simply as an event that conforms to a pattern that can be determined independently of the event. A pattern representing a specified event is, in turn, called a specification. For example, the faces of Washington, Jefferson, Roosevelt and Lincoln are specifications with respect to certain rock formations on Mt. Rushmore. Specifications do not have to be defined in advance, as long as the pattern can be constructed without knowledge of the event in question.

Dembski's universal probability bound is based on three facts: 1) the number of elementary particles in the universe: 10^{80} particles, 2) the maximum rate at which transitions in physical states can occur (inverse of Planck time[1]): 10^{45} transitions per second, and 3) the age of the universe: 10^{25} seconds[2]. Dembski (2004) deduces that because every specified event

[1] Planck time (5.4×10^{-44} seconds) is the shortest unit of time that can be observed in physics. It is defined as the time that is required of a photon travelling at the speed of light to travel a distance of a Planck length (the shortest possible observable length) in a vacuum.

[2] According to current scientific understanding, the universe is actually a billion times younger than this, ca. 10^{17} seconds. Dembski seems to want to play it safe here.

requires at least one elementary particle to specify it, and because such specifications cannot be generated faster than Planck time, the number of specified events through the history of the universe must fall below 10^{150}. Thus, every specified event whose probability is less than the universal probability bound is highly improbable. Dembski points out (rightly, actually) that his probability bound is the most conservative in the literature.

Dembski considers the consequences of his theory to be of extreme significance. Firstly, if complex specified phenomena can be detected in the biological world, there has to exist a supernatural intelligent designer who has designed these phenomena. Secondly, since methodological naturalism is incapable of explaining design, it must be abandoned as a basis for science and replaced by ID. For Dembski, the theory of intelligent design is "a bridge between science and theology". In the light of the theory, many features of the physical universe can be interpreted as evidence of the creative work of an intelligent designer. In the context of Christian theology, of course, the designer is usually identified as God.

Necessity, Chance, and Design

Dembski's theory has attracted a lot of criticism. It has been suggested that the notion of specified complexity is not defined rigorously enough to constitute a reliable basis for detecting design. For instance, the subjectivity of defining specifications, the questionability of the universal probability bound, and difficulty of defining specifications independent of probabilities, are claimed to pose serious challenges to the theory of intelligent design (e.g., Sober, 2002; Bartholomew, 2008; Doran, 2010; Elsberry & Shallit, 2011). In this paper, I focus on Dembski's three modes of explanation of events – necessity, chance, and design – which he claims to be mutually exclusive and exhaustive. My main point of criticism is that Dembski does not give any logical reasons for why his trichotomy of explanatory modes would hold true. He simply appeals to our everyday intuition that all events around us are caused by either necessity, chance or design (Dembski, 1998). My intuition goes against Dembski's here. I think that combinations of these three modes might well be possible.

Consider one of Dembski's favourite examples, that of an archer shooting an arrow at a target. Dembski pictures a situation where an archer hits a fixed target from 50 meters away. In this case, the event of hitting a target is specified and we should attribute it to design. Or should we? Certainly, there is design involved in the archer aiming the arrow and giving it a certain velocity. After that, however, the flight of the arrow is governed by the laws of physics – that is, by necessity. Perhaps chance may be involved as well, for example in the form of a blast of wind. Thus, design, necessity and chance can work together. Undoubtedly, this example is extremely simplified; but it is one that Dembski uses, so it is appropriate to use it in turn to criticise his theory.

Or, let us take another perspective and consider an event E which is a disjunction of subevents $E_1, \ldots, E_n$. Now, there can exist an E which is too improbable not to be designed even though all subevents are probable enough to be designed. Since Dembski's argument will, in this case, lead to different conclusions depending on whether the event is examined as a whole or as a combination of its parts, it is impossible to decide whether E is really designed or not. A similar example can be constructed when an event E (which is allegedly designed) is a conjunction of subevents $E_1, \ldots, E_m$ (which can still be probable enough to be caused by chance), and also when an event E is a conjunction of a designed event E_1 and chance event E_2 (Fitelson, Stephens & Sober, 1999).

All in all, separating design, necessity and chance as three independent phenomena simply does not work – or at least Dembski needs to give us better reasons than just "everyday intuition" for accepting this policy. I would rather be inclined to think that every event occurring in the universe is an outcome of complex intertwined processes characterised by a combination of necessity and chance, and – possibly – design.

ID's Argument Against Methodological Naturalism

In biology, the evolution of living organisms can be described by the Darwinian mechanism of natural selection and random mutations, which corresponds to the interplay between necessity and chance. ID advocates hold, however, that there are some biological organisms that cannot have been brought about by necessity and chance, i.e. natural causes. These biological entities are claimed to be too complex and too information-rich to have emerged by natural causes, and can only have been produced by a supernatural intelligent agent.

The most famous example – and, according to many critics, the only one even to some extent credible example – of a biological organism that is allegedly designed is the *Escherichia coli* bacterium (which has lately become some sort of a mascot of the ID movement). According to biologist Michael Behe, the flagellum is "irreducibly complex," meaning that it could not have been formed by numerous, successive, slight modifications. Dembski, for his part, sees the flagellum expressing specified complexity and, thus, design. Recently, these views have been refuted by many scientists (e.g., Pallen & Matzke, 2006; Van Till, 2003; Wong et al., 2007), but the flagellum example still seems to persist.

Methodological naturalism, instead, rules out any kind of design or divine action in nature by definition. Clearly, this contradicts ID at a fundamental level. The argument of ID proponents against methodological naturalism can be put as follows:

1) If design inferences concerning some natural phenomena are warranted, then an intelligent agent has brought about such phenomena.
2) If an intelligent agent has brought about some natural phenomena, then natural causes will fail to explain such phenomena.
3) If natural causes fail to explain certain natural phenomena, then methodological naturalism will in some cases lead to errant explanations.
4) Any methodology that leads to errant explanations should be abandoned.
5) Design inferences concerning some natural phenomena are warranted.
6) Therefore, methodological naturalism should be abandoned.[3]

This is certainly a strong claim, and also a rather logical one. If there is a methodology that does not work, should it not be abandoned?

Are Intelligent Design and Methodological Naturalism Really Not Compatible?

I argue that there are some problems with the preceding argument. Certainly, premise 5 is highly dubious. The issue of whether any of the design inferences suggested by the ID movement are warranted, however, is beyond the scope of this paper. But let us assume, for the sake of argument, that there do actually exist some natural phenomena that can be inferred as designed. In this case, should we reject methodological naturalism? ID's answer is "yes," mine is "no."

ID holds that detecting design always implies supernatural intervention, meaning that the laws of nature need to be overridden in order for specified complexity to appear. But why should we accept such a

[3] I am here paraphrasing, and extending, the formulation presented by Murray (2006), who has revealed some shortcomings in the argumentation of ID advocates.

claim? I see no reason. I believe that we do not need to set natural causes and design against each other. Just as design, necessity and chance worked together in the archer example, they can work together in all events. My point is that although some biological objects may not have been brought about by natural causes alone, they are still biological objects and have been brought about partly by natural causes. Thus, the second premise is false and the argument against methodological naturalism fails.

Of course, the scope of naturalistic explanations is always limited since, by definition, they disregard the possible supernatural component. So methodological naturalism may offer correct, but potentially partial, explanations. I think the argument against methodological naturalism should be refined accordingly, in such a way that methodological naturalism should be required to be complemented with further explanations:

1) If design inferences concerning some natural phenomena are warranted, then an intelligent agent has brought about such phenomena.
2) If an intelligent agent has brought about some natural phenomena, then natural causes will fail *fully* to explain such phenomena.
3) If natural causes will fail *fully* to explain all natural phenomena, then methodological naturalism will in some cases lead to *partial* explanations.
4) Any methodology that leads to *partial* explanations should be *complemented with further explanations.*
5) Design inferences concerning some natural phenomena are warranted.
6) Therefore, methodological naturalism should be *complemented with a theory that explains design.*

Still, methodological naturalism should not be abandoned. After all, it is a methodology that has enabled scientists to make amazing discoveries during the centuries. ID, in turn, has not provided an alternative positive research programme. We cannot predict in advance what some specific features of the world would look like if they were designed, because we cannot know the intentions of the supernatural designer. In my opinion, this is one of the biggest problems of ID. Design cannot be predicted, laws of nature can. Therefore, even if scientists were someday, somehow, to reach consensus on some particular natural object being designed, they would still have to go back to using methodological naturalism in their everyday research, by definition (Collins, 2006). At the moment, methodological naturalism is really "the only game in town", for science.

New Vision for ID

It seems that the ID movement needs a new vision. The current theory, which claims to be a scientific theory, simply leaks from too many sides. In this paper, I have only scratched the surface. Further problems are concerned with Dembski's treatment of probability theory, as well as his understanding of formal logic, which are argued to be incomplete. The views of other main proponents of ID, such as biologist Michael Behe (1996; 2007) and philosopher Stephen C. Meyer (2009; 2013), have also received substantial criticism. To be honest, I am not quite sure what exactly the new vision for ID should look like, but at least it should not be based on the artificial trichotomy of explanations of events, nor on the claim that God needs to violate the laws of nature to act in the world.

One approach that might be fruitful is that of information theory. Perhaps God can be seen as an "information inputter" who controls events in the world by entering information at convenient times and places. Thoughts of this kind have been presented by, for example, John Polkinghorne (1989), who sees God as an inputter of "active information" – and this is actually what Dembski asserts in many of his writings. According to Dembski (2002), objects that express specified complexity actually reveal complex specified information that a supernatural designer has entered into the world. However, Dembski's theory does not allow any input of information without bypassing the laws of nature and moving particles in some "unnatural" way, so it leads to the same conflict between design and natural causes as discussed above. I believe that, if Dembski developed his theory of information input in a way that did not require divine intervention, it might become one of the more fruitful aspects of the ID case. The information-theoretic perspective of ID is, however, beyond the scope of this article.

Having criticised ID in this paper, I still do not think we should condemn it as a complete failure – at least if it is understood in a slightly different way than usually. If ID could be seen as a philosophical view asserting that events occurring in the world are correctly, but partially, described by natural causes, and therefore correctly but partly explained by methodological naturalism, ID would be compatible with methodological naturalism. And if ID could be seen as a theological standpoint defending the basic claim that any complete explanation regarding the world should take into account possible supernatural causes, supernatural intelligent designers – perhaps one would even dare to say "God" – ID would be in harmony with mainstream Christian theology. This way ID could contribute more fruitfully to the debate between science and religion – and to the search for an Intelligent Designer.

References

Bartholomew, D. (2008) *God, Chance and Purpose. Can God Have It Both Ways?* Cambridge: Cambridge University Press.

Behe, M.J. (1996) *Darwin's Black Box. The Biochemical Challenge to Evolution*. New York: The Free Press.

—. (2007) *The Edge of Evolution. The Search for the Limits of Darwinism*. New York: Free Press.

Collins, R. (2006) "A Critical Evaluation of the Intelligent Design Program" (http://home.messiah.edu/~rcollins/Intelligent%20Design/INTELL3.htm).

Dembski, W.A. (1998) *The Design Inference. Eliminating Chance through Small Probabilities*. Cambridge: Cambridge University Press.

Dembski, W.A. (2002) *No Free Lunch. Why Specified Complexity Cannot be Purchased without Intelligence*. Lanham, MD: Rowman & Littlefield.

—. (2004) *The Design Revolution. Answering the Toughest Questions about Intelligent Design*. Downers Grove, IL: Inter-Varsity Press.

Discovery Institute, Center for Science and Culture (2014) *Intelligent Design FAQs*. Available at: http://www.discovery.org/id/faqs/#questionsAboutIntelligentDesign. Accessed 28 August 2014.

Doran, C. (2010), "Intelligent Design. It's Just Too Good to Be True" – *Theology and Science* 8(2), 223–237.

Elsberry, W. & Shallit, J. (2011) "Information Theory, Evolutionary Computation, and Dembski's Complex Specified Information". *Synthese* 178(2), 237–270.

Fitelson, B., Stephens, C. & Sober, E. (1999), "How Not to Detect Design – Critical Notice: William A. Dembski, "The Design Inference" – *Philosophy of Science* 66(3), 472–488.

Meyer, S. C. (2009), *Signature in the Cell. DNA and the Evidence for Intelligent Design*. New York: HarperOne.

—. (2013), *Darwin's Doubt. The Explosive Origin of Animal Life and the Case for Intelligent Design*. New York: HarperOne.

Murray, M. J. (2006), "Natural Providence (Or Design Trouble)" in *Philosophy of Religion: An Anthology*, 6th Edn, ed. L.P. Pojman & M.C. Rea. Belmont, CA: Wadsworth. 596–612.

Pallen, M. . & Matzke, N. J. (2006), "From The Origin of Species to the Origin of Bacterial Flagella" – *Nature Reviews of Microbiology* 4(10), 784–790.

Polkingthorne, J. (1989), *Science and Providence. God's Interaction with the World*. London: SPCK.

Sober, E. (2002), "Intelligent Design and Probability Reasoning". *International Journal of Philosophy of Religion* 52(2), 65–80.

Van Till, H. (2003), "Are Bacterial Flagella Intelligently Designed? Reflections on the Rhetoric of the Modern ID Movement". *Science and Christian Belief* 15(2), 117–140.

Wong, T., Amidi, A., Dodds, A., Siddiqi, S., Wang, J., Yep, T., Tamang, D.G. & Saier, M.H. Jr. (2007), "Evolution of the Bacterial Flagellum". *Microbe* 2(7), 335–340.

CHAPTER TEN

FINE-TUNING ARGUMENTS FOR THE EXISTENCE OF GOD: A SHOT IN THE FOOT?

RICHARD GUNTON

Introduction

Fine-tuning arguments try to persuade non-theists that there may be a supernatural designer of the Universe. They are a type of design argument that draws upon the science of physics (other design arguments draw more upon chemistry and biology for example). In this short article I begin by outlining the structure of fine-tuning arguments, then mention a problem with the derivation of the tiny probabilities sometimes invoked and then consider the development of the "Anthropic Principle". This prepares the ground for my main focus, which is a more radical critique concerning thought experiments, possible worlds and multiverses.

What are fine-tuning arguments? First, here is geophysicist Bob White on www.rejesus.co.uk:

> The universe is extremely finely balanced. The anthropic principle, which is a subject of scientific study, is concerned with some of these incredibly precisely tuned parameters that make life possible. Someone once suggested that the tuning of some of the physical constants is as accurate as hitting a one centimetre square target with an arrow fired from the other side of the universe.

This is certainly a graphic illustration – and a very strange idea. Where does it come from? Here is something from the apologetics site www.bethinking.org:

> The gravitational force must be what it is for planets to have stable orbits around the sun.... It is estimated that a change in gravity by only one part in 10 to the power of 100 would have prevented a life permitting universe. … Not only must each of these quantities be exquisitely fine tuned but their ratios to each other must be finely tuned. As William Craig writes, "Improbability is added to improbability until our minds are reeling in incomprehensible numbers."

We see here that, although physical constants are normally assumed to be *invariable*, fine-tuning arguments encourage us to imagine them having different values. Now, all measurements are made with some amount of *error*, so one can easily imagine the best-known value to be *wrong* by a given amount within some range of uncertainty. But that is not what is going on here. Here accepted laws of physics are being used to explore how much certain physical constants *could vary*, in the models, from their assumed values without preventing the modelled universe from being conducive to the eventual appearance of us, the "observers". In this way, certain tolerance levels are found. And they typically show that the Universe as (we think) we know it inhabits a very tiny island in a sea of otherwise barren parameter space. In a nutshell: "If the structure of the Universe were slightly different, we would not exist." Some people seem to be deeply impressed by such a claim; others (myself included) do not. In the rest of this article I explore some points at which people who are impressed by fine-tuning arguments may have been misled by poor argumentation. But first we look at some empirical statistics.

Do fine-tuning arguments work?

A couple of recent surveys show something interesting. First, Helen De Cruz polled 802 philosophers and found that *the argument from design* (which includes fine-tuning as well as biological design arguments) showed relatively high credence by both theists and atheists, but with one of the greatest discrepancies between these two groups: theists found it about twice as convincing as atheists did (De Cruz 2014). Then, in a survey of 210 scientists that I conducted (Gunton *et al.*, in prep.), *fine-tuning* arguments won more approval than other design arguments from Christians, and less from those respondents classified as "Modernists" (Fig. 10-1). Both these surveys suggest that people's religious position determines how much they approve of such arguments – a point that I will return to in closing.

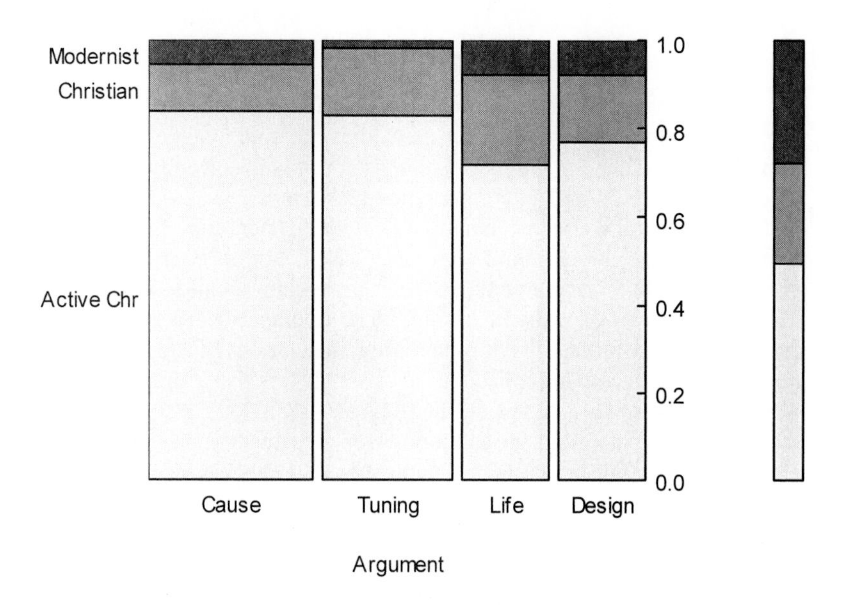

Fig. 10-1. Religious positions of 160 respondents who "agreed" or "tended to agree" that one or more of four arguments for the existence of a deity are compelling. The arguments were: (1) Cause = "The laws of nature or the origin of the Universe require an explanation that is outside the Universe;" (2) Tuning = "The improbability of physical constants being conducive to a habitable Universe by chance calls for an explanation outside of nature;" (3) Life = "Living things are so different from non-living things that they could never arise by natural processes from chemicals;" (4) Design = "The intricate structures or functions of (at least some) biological systems could never arise by natural laws." "Modernist" was defined as any of atheist, agnostic, (Christian) humanist or "none", while "Active Christian" was attributed where a respondent indicated "Christian" for their religious affinity, belonging to a religious community and holding to an established belief system. The bar to the right indicates the proportions of all 179 respondents who fell into one of the three religious categories: 28% Modernist, 22% Christian, 50% Active Christian.

I cannot deny that such arguments do have persuasive power for some people. I do not, however, think that they would, hypothetically, persuade me not to be an atheist.

A technical problem

The tightly-defined island of parameter space mentioned above is often turned into a probability value, or used to evoke some vague idea of vast improbability. Take the gravitational constant: this is currently believed to be 6.67384×10^{-11} N m^2 kg^{-2} ("Gravitational constant", *Wikipedia*, accessed 16 Jan 2015), and we saw a claim that if it were different by 1 part in 10^{100}, a life-permitting universe could not have arisen. But how can we obtain a *probability* about the gravitational constant taking a value conducive to life "by chance" (i.e. without design)? The tolerance value of 1 in 10^{100} certainly isn't a probability; probabilities must be estimated either from frequencies of events in replicate trials – which we clearly don't have here – or from assuming a sampling distribution – and we have no basis for assuming any sampling distributions for constants. The problem of obtaining probabilities from uncharacterised sample spaces has been carefully argued by McGrew *et al.* (2001), and is not the focus of this article.

Anthropic Principles and Perspectives

Some presentations, like that of John Polkinghorne (2007), avoid invoking specific probability values, relying instead on a vaguely evoked sense of astronomical improbability. But then we often find some appeal to "the Anthropic Principle". A principle by this name is in fact invoked by both design protagonists and their sceptics. This is therefore an important issue to clarify, and forms a bridge to my main argument.

The Anthropic Principle (AP) was initially introduced by Brandon Carter (1974) in opposition to the Copernican Principle, which stated that we do *not* occupy any special location in the Universe. Carter's "weak" AP stated that humanity can only find itself at such a time and place in the Universe as the laws of nature make conducive to our existence. Note that this is meant as an *a priori* truth about us as observers with respect to the Universe: we do not need any particular experience in order to be sure that the nature of the Universe is conducive to our existence. Carter also offered a "strong" AP, which states that the constants of physics must also be conducive to our existence: a similar *a priori* claim about the Universe with respect to us. Some confusion was then introduced when a 1986 book "*The Anthropic Cosmological Principle*" (Barrow & Tipler, 1986) defined APs differently. And this seems to have started an awkward tradition of redefining the AP as a reference to fine-tuning. For example, Polkinghorne (2007, p .1) says, with uncharacteristic obscurity, "The central insight of the … AP is that the specific character of lawful necessity had to have a very particular form … if the coming-to-be of *anthropoi* were to be possible …" Such a statement does not have any of the *a priori* force of the original AP (in either its strong or its weak form); it sounds more like a re-statement of the fine-tuning thesis. Such an "anthropic principle" seems to set up our own existence as a possibility rather than a given – perhaps an attempt at a "God's-eye view" of the early Universe.

I try to shed more light on this below. For now, I'd propose that we retain the strong AP as a simple *a priori* truth: given that we exist, we can know that the Universe *must* be suitable for our existence (and we should not expect anyone to be surprised at this). This AP may also be taken as the conclusion to a premise derived from an act of *reflexive* reference, because the speaker takes his existence as given. In this form it corresponds rather closely to Descartes' famous *cogito*, and indeed Carter stated the strong AP analogously as, "*Cogito ergo mundus talis est*" (I think, therefore the world is such as it is). Specifying just *how* the Universe is conducive to our existence is not part of the claim offered as an *a priori* truth; this question may be an intriguing subject for physical modelling, and indeed the tiny island of hypothetical parameter space mentioned above may constitute one of the most interesting answers to this question. But in its basic, original form, as an *a priori* truth, I suggest that the AP remains fit to undermine fine-tuning arguments by removing the element of surprise on which they depend.[1]

To summarise so far: I argue that fine-tuning arguments are non-starters because (1) we can't put probability values on cosmological constants being suitable, and (2) any argument that appeals to surprise at the Universe's habitability can also be defeated by the Anthropic Principle when properly construed. Next I argue that (3) any attempt to construe the Universe as an instance of a physical process is thoroughly non-empirical, and indeed (4) begs the question of "divinity" in a general sense.

[1] This can also be explained in terms of the Bayesian calculus of probabilities: if we take our own existence as a given (as I think we must), then any *prior* improbabilities in the conditions we observe must be factored into calculations of the *conditional* probability of our observing those conditions *given that we exist*, so that the improbabilities cancel out. (The illustration of a firing squad is sometimes used at this point - but the analogy is problematic, when we think about the prior beliefs in a Bayesian way.)

On thought experiments and possible worlds

Think for a moment about how we use the word "if" in counterfactual propositions. "*If* I had run, I *would have* caught the bus." Or, more profoundly, "…if the miracles that were performed in you [Capernaum] had been performed in Sodom, it would have remained to this day" (Matthew 11:23). There is no way to inquire further about whether these propositions are true, because they refer to imaginary situations that simply don't exist (I *didn't* run; the miracles *weren't* performed). But clearly we still find such claims useful. I suggest this is because they describe regularities or laws: they express certain ways of understanding the world, world pictures where one thing leads to another. I may be claiming that running *generally* enables me to reach bus stops earlier than walking, and Jesus *characterises* the people of Capernaum in contrast to those of Sodom in a certain way. We may refer to "possible worlds", and thought experiments.

Counterfactual "if" propositions, then, seem to be rhetorical tools enabling a speaker to communicate a way of understanding the world by calling on listeners to imagine a situation and some consequences. But sometimes we encounter a more ambiguous "if" sentence. Maybe we've been in a situation where members of a group are encouraged to introduced themselves to each other with a question like, "*If* you were an animal, what would you be?" Here we should recognise that we're being asked to share something of our character by drawing on folk zoology. But some thought experiments may stop us in our tracks. A counterfactual "if" clause may call into question one's basic beliefs, parts of one's worldview. Something like "If I were a tree, I would have a more peaceful life" might do this for many people. Can I perform this thought experiment? Do I believe in any laws or principles that enable me to imagine consequences for this outlandish counterfactual?

So, back to fine-tuning arguments. "If the physical structure of the Universe were slightly different, we would not exist." Faced with this exotic claim, and perhaps before appealing to anthropic principles, we ought to consider what kind of laws or principles are being invoked here, such that the "if" clause will entail the consecutive clause. Try this one: "If *pi* were even slightly different from what it is, then circles would not be round!" Well – who says? The nub of the matter is that fine-tuning arguments ask us to imagine the constants of physics being *different* while the relationships between them *remain* as enshrined in currently-accepted physical laws. And this suggests that laws of physics are taken to transcend the Universe as we know it, while (at least one) constant is not.

So this counterfactual thought experiment depends upon believing in a higher level of law than anything from the Universe that we know – a law for *universes in general*, we might say. Some people may have such a belief, but it sounds distinctly metaphysical – and, I will now suggest, not very Christian.

A divine multiverse?

Fine-tuning protagonists generally reject the idea of a universe of universes, so to speak, as just such a metaphysical belief. It should come as no surprise that I agree here. I would even call this idea "religious", in the absence of any belief in a creator God, because it seems to posit an ultimate reality that can explain what we experience. But is there not an irony here, if fine-tuning arguments also assume a reality for physical laws at a level beyond the Universe? We might say that fine-tuning protagonists posit a "multiverse" of possible worlds, just as some of their opponents posit an *actual* multiverse. The fine-tuning protagonist imagines an infinite "multiverse" of possible worlds and urges amazement that the *actual world* lies in the inhabitable "sweet spot"; his opponent imagines an endless succession of universes in an *actual multiverse* and so is *un*surprised to find himself in one that happens to be inhabitable. This analogy may suggest some new lines of dialogue between theists and naturalists. But what seems clear is that the question of what physical laws hold in either of these imaginary realms is not an empirical one; we may say that it is up to the whim of each person in the discussion. And this kind of metaphysics will reflect people's religious beliefs.

Why do I invoke "religious" beliefs here? Roy Clouser (2005) argues that all theories implicitly take something or other as *divine*, where "divine" refers to whatever is held to be "unconditionally non-dependent reality" – that is, as self-existent (eternal and uncreated) and giving rise to everything else. This definition follows most of the pre-Socratic philosophers and Plato as well as a number of more recent philosophers (Clouser, 2005). Now, if fine-tuning arguments work by assuming a reality for physical laws beyond the observable Universe, then even if they are successful, we may suspect that they set up physical laws to vie for the status of divinity with the "God" supposedly inferred. To put it another way: arguments colour the conclusions they produce. If fine-tuning arguments succeed in persuading someone, the kind of "God" they most naturally suggest is a "divine fine-tuner" (in the words of Polkinghorne, 2007: p.3). But such a deity appears to perform fine-tuning within the *absolute* constraints of a law-like cosmic dashboard, and to dislodge these constraints or laws from the status of divinity may be a pernicious challenge. And whether or not the arguments succeed, at the very least they risk tarring their protagonist with belief in a competing divinity, in the general sense of "something absolute".

Conclusion

My driving concern has been that fine-tuning arguments carry too much atheistic metaphysical baggage to be effective tools in an apologetic for monotheistic faith. At the outset I said that if I were an atheist, I don't think I'd be persuaded by fine-tuning arguments. I used this counterfactual "if" statement to make a claim about the rationality of such arguments. The possibility that I might actually be, or become, an atheist is irrelevant here; indeed from my own perspective it seems quite remote. Having said that, it doesn't seem as remote as the idea of the non-existence of that particular creator God revealed through the Jewish tradition and in Jesus of Nazareth.

References

Barrow JD and Tipler FJ (1986) *The Anthropic Cosmological Principle.* Oxford University Press.

Bertola F and Curi U (1993) *The Anthropic Principle.* Cambridge University Press.

Carter B (1974). Large Number Coincidences and the Anthropic Principle in Cosmology. *IAU Symposium 63: Confrontation of Cosmological Theories with Observational Data.* Dordrecht: Reidel. pp. 291–298 (cited on Wikipedia, "Anthropic Principle", 11 March 2012).

Clouser R (2005) *The Myth of Religious Neutrality: An Essay on the Hidden Role of Beliefs in Theories* (2nd edn). University of Notre Dame Press.

De Cruz H (2014). The enduring appeal of natural theological arguments. *Philosophy Compass* 9: 145–153.

Gunton RM (in prep.) A survey of scientists' worldviews.

Leslie J (1989) *Universes.* New York: Routledge.

McGrew T, McGrew L, Vestrup E (2001) Probabilities and the Fine-Tuning Argument: A Sceptical View. *Mind* 110: 1027-1037.

Polkinghorne J (2007) *The Anthropic Pinciple and the Science and Religion Debate.* Faraday Paper No. 4, The Faraday Institute for Science and Religion, Cambridge.

CHAPTER ELEVEN

DUALIST PERCEPTIONS AND PITFALLS

JOHN EMMETT

Introduction

This chapter asks whether the duplet in the Conference title, "Laws of Nature, Laws of God?", though arguably innocuous in itself, is not an example of a general, and often more dangerous tendency. It is suggested that using duplets such as "God and Nature" risks evoking dualisms such as "supernatural and natural", "spiritual and physical", even "religion and science". The perceptions and pitfalls, dualist traps, of such descriptions are considered, and so is how they may be avoided. In the Conference discussions, a new duplet was recognised; the question whether the laws that frame existence are prescriptions or descriptions? That, too, is considered in what follows.

The Perceptions and Pitfalls

What, then, are these dualist perceptions and pitfalls? The perceptions are the dualist tendencies of "popular" theology that separate God and the world, the spiritual and the physical, the supernatural and the natural. The theology is "popular" because it is spoken and heard by people at large, whether they attend Christian worship or not. This is a real challenge for the Christian Church to provide good apologetics, an account of the Christian religion that engages with contemporary culture. The pitfalls are similar, in that the fundamental dogma of Christianity, God as Trinity, is not taken seriously enough. There is a tendency in Christian teaching and preaching to avoid the subject because it is too much of a conundrum, too academic, too theoretical, not practical enough, and not "Biblical". This again is a challenge for the Church. The Trinity, the effective hallmark of Christianity, discernible in the Bible but forged into a dogma through subsequent reflections by the theologians of the Early Church, seems to sit uneasily with Christian practitioners. So the Church must provide good dogmatics and systematics, a coherent, comprehensive and consistent account of the tenets of the Christian religion, one that engages with contemporary knowledge, and especially in this context with scientific method and discovery.

The Approach

The argument of this paper is that the way in which the popular dualist traps can be avoided is to deploy a Trinitarian model of God that works strictly with threesomes and not twosomes. The model to be used here originated with St Augustine of Hippo, who described the Spirit as the bond between the Father and the Son (Augustine [339-419], 1991). The model also draws on the insights of panentheism, process, and kenotic theology. The way in which it is presented is through a series of simple diagrams that depict flows between Father and Son, heaven and earth, the eternal and the temporal. Thus:

Father <= Spirit => Son
heaven <= space => earth
the eternal <= time => the temporal,

where the arrows represent an out-flow, cross-flow and in-flow from one side of the triadic pattern to the other. In order to depict the triune God there has to be a balanced contra-flow, no one side dominating the other. This approach (Emmett 2001) has been developed into a way of describing the basic tenets of the Christian religion using ideas that have a similarity with the concepts of quantum physics (Polkinghorne, 2007). The model can provide a Trinitarian analysis of a variety of themes (Emmett, 2007).

The Dualist Trap

What then are the dualist traps that need to be avoided? This can be depicted as a series of one-way flows

God => influencing => Nature
the supernatural => influencing => the natural
the spiritual => inspiring => the physical
creator => creating => the created
creator => ordering => cosmos
designer => designing => the designed.

The most important point here is that the word "God" appears on the spiritual and supernatural side, rather than in the Trinitarian model where God is the whole relational pattern of Father, Spirit and Son. It is a common manifestation of dualist traps, that the triune God is equated with the First Person of the Trinity, Father in the New Testament or Lord in the Old. God is a "He" rather than, as triune, a relationship for which there is no really helpful pronoun. Another important point is that there is no reverse flow, the creator God is not influenced by the natural and the created. This preserves the "impassibility of God", but the price is a simple model of an intervening heavenly God and a fatalistic perception of earthly, human life.

The Trinitarian Model

The Trinitarian model used here requires a two-way flow for the bond of the Spirit to make the relationship between the Father and the Son:

Father <= Spirit => Son

Again, following St Augustine, the bond of the Spirit is the bond of love ..

lover <= love => the beloved

.. and there are other triadic patterns that can depict the triune God in this way. Each one describes, according to this model, some aspect or attribute of the triune God, the God of the Christian religion. Remember that God, as Trinity, is the whole pattern and not in any one position in the triad as in the above dualistic patterns. Thus, the spiritual is placed in the central position of a general triadic pattern ..

the non-physical <= the spiritual => the physical

.. where the physical and the rather clumsily labelled "non-physical" are a complementary pairing that in general also represents several others such as infinite-finite, invisible-visible, ineffable-effable, immortal-mortal. However all are connected by "the spiritual" just as "Spirit" is the connective bond of the classical three Person triad of the Trinity. Another triad, which avoids gendered words, is ..

deity <= life => humanity

… where “life” refers to the spiritual life rather than simply life on earth. The same pattern could be applied to other theistic faiths if the religious life is for humanity a connection with deity, a spiritual life, even though the name of the deity and the understanding of the connection may differ between faiths. Could this be a useful point perhaps for a dialogue between faiths?

These general triads have important implications. The triune God now contains the physical universe, the created order, the world and humanity. In general this can be depicted as ..

the transcendent <= the connective => the immanent

… which has a similarity to panentheism (Clayton & Peacocke, 2004) where God is both with-out, as transcendent, and with-in, as immanent. God is not just with-out, beyond, as in the dualist trap mentioned earlier, nor just with-in as in pantheism. In relation to the universe, the world, nature, humanity – however aspects of the immanent are described – the triune God, according to this model, is with-out, with and with-in all space and time of existence. The concept of God within the world, and within humanity in particular, is at the heart of the Christian religion because, in and through Jesus, the Christ, the divine became human and thus all humanity participates in the life of the triune God.

Each of these triadic patterns depicts an aspect of the inner essence of the triune God, sometimes called *the immanent Trinity*, or better here, to avoid confusion, *the essential Trinity*. An important triad that depicts the outer work of the triune God is:

creator <= sustainer => redeemer

This has been called *the economic Trinity* because it is an attribute of how the triune God operates, without, with and within the natural world. The one-way flow of creating was depicted earlier when describing the dualist trap. For this flow to participate in the life of the triune God it must be combined with a reverse flow of redeeming:

the redeemed <= redeeming <= redeemer

The created order, according to Christian theology, is a work of love and gift. The triad pattern for love was introduced earlier. For gift the triad is:

giver <= gift => receiver

Note that both love and gift involve a two-way flow, otherwise they are just accepted without any response.

The triadic patterns of both the essential and the economic Trinity will be important when applied to the Conference topic of Laws of God and Nature. However before that, two other important requirements of any Trinitarian model need to be mentioned. One is *perichoresis,* translated here as "co-inherence". The three Persons of the Trinity are with-in each other in both essence and economy, and so this must apply to all the triadic patterns of this model.

The word "flow", describing the connecting of the triads, is deliberately chosen to denote a dynamic and indeterminate bond. The Spirit has a freedom but also a direction, guiding and carrying a source to a goal in a playful dance – which is how the word *perichoresis* is sometimes interpreted. Interestingly, there are similarities between this model and the way quantum physics models the fundamental forces of sub-atomic particle interaction in the dynamic, indeterminate "dance" of virtual particles and the exchange of a "real" particle (boson) to connect two other particles (fermions).

Laws of Nature, Laws of God

Having posed the problem – the dualist pitfalls and perceptions – the analytic approach of the Trinitarian model can now be applied to the question in the Conference title: "Laws of Nature, Laws of God?" The laws of the natural sciences rather inevitably raise the question of creation. Are the laws just part of the created order, or are they perhaps the direct work of a creator, a deity of some kind? The Trinitarian model demands that it is not a question of either/or but both/and. The panentheistic approach has both the creator and the created, the designer and the designed within the life of (the triune) God. So God is not only with-out the natural world, creating its laws, but also with-in the created laws of the world themselves.

God contains both the transcendent and immanent as Father and Son respectively. The connective Spirit not only binds these two together but also opens up a "space", a separation and a letting-go, between creator and created. This means, for the created order, a letting-be and letting-do, a dynamic, uncertain, freedom. To create in love is to let the beloved be and do. This opening of the "space" within the life of the triune God is how the Biblical concept of *kenosis,* describing the incarnation of Jesus the Christ, can be applied to the whole of the created order, the universe, the world, humanity (Polkinghorne, 2001).

Process

The contra-flow, sustaining existence in and through creating and redeeming, can be depicted by other general triads. One has similarities to process theology (Cobb and Griffin 1976):

the potential <= process => the actual

The process of existence in space and time can be understood as a flow of potential to actual. What happens in the actual here and now is the effect of many potential causes from the past there and then. Once the event has happened it becomes a potential for the future there and then. These connections between potential and actual are a way of depicting the web of cause and effect of existence, and the flow called the "arrow of time", from past through present to future.

The work of science as a human endeavour is to try to understand the process of existence. This is usually done with the aid of conceptual models which, in the physical sciences, are expressed in mathematical formulae. These are essentially trying to relate potential and actual. If the formula gives a reasonable description of experimental observation, the theory is accorded the status of a law, is taken on trust by the scientific community, and features in science textbooks. From then on it is taken for granted as a given. Even so, the law is neither perfect nor absolute, because there may be situations where a very different model is needed. So why should such laws and theories work for science, and why, of the different potentials, does one describe the actual better than any other?

The scientist, working in the actual world, observing actual data, is trying to describe which potential has become actual. Science takes the actual as a given, a gift; but who or what, if anything, is the giver? The Christian theologian would claim that the giver is none other than the creator, Father God. By contrast, the atheist would perhaps claim that the actual is *just* a given, and that's that. Does the creator prescribe which potential becomes actual? In a dualist sense the laws of nature are simply prescribed, written out beforehand, by the creator God for humanity to discover, describe, write down. The Trinitarian model suggests that the laws of nature are both prescribed and described, written out *and* written down. There is no sense of beforehand either, only a continuous interplay in space and time, of potential past and future with actual present. Both deity and humanity, creator and the created, co-operate in the understanding of the laws of existence

This approach has some interesting similarities with what is still sometimes termed "modern physics", the theories of relativity and quantum mechanics. The idea of a particular potential becoming actual is similar to two interpretations in quantum theory, either the "Copenhagen interpretation" of the collapse of the wave-function, or the multiverse interpretation of the universe being one of a potentially infinite number. The idea that there is an interplay of the prescribed and described, the work of the creator and the created, is similar again to the Copenhagen interpretation, that a potential prescription requires an actual description. For a potential observation to become actual requires the operation of an observer.

The Trinitarian model avoids the dualist trap by not defining an active God influencing a passive natural world but rather a God of interplay, of prescribing and describing, the work of deity and humanity respectively. The laws of nature are prescribed, or preferably proscribed, having their source in the transcendent and are given to the immanent for humanity to describe. This understanding of existence is all contained within the life of the triune God. The scientist who tries to describe these laws is participating in this life of the triune God. So there is not a dualist division between the laws of God and Laws of Nature; rather the Laws of Nature are the laws of God, and vice versa, provided the concept of God is a fully Trinitarian one. The process of prescribing and defining the laws of nature is a dynamic interplay between the observer and the observation that is contained within the life of the triune God.

Purpose

It has to be recognised that the atheist would probably maintain that the process of nature is just given, without any reference to a connection with a transcendent deity who lovingly gives the created order, including the human beings who observe it. However, alongside the given and received process of the created order there is also the question of whether the process carries with it a purpose. This introduces another general triad that depicts the triune God:

the potential <= purpose => the actual

Note that both process and purpose connect the potential and actual. Of all the potentials why this particular actual, and what is the potential purpose of this actual? Again it has to be recognised that the atheist would probably maintain that existence has no purpose or that its purpose lies within itself without any reference to a deity who gives meaning to the created order. However that is not the observation of the Christian theologian. The dynamic interplay of potential and actual within the life of the triune God contain both process and purpose. In general, science primarily asks questions about process, and religion asks questions about purpose. Although it is possible to pursue science without any religious faith, once questions of purpose are raised, not least in how scientific advances are used, then religion can rightly claim to have its voice heard. That claim is made here on the basis that both process and purpose are contained in the life of the triune God.

Dogmas

Religion, too, has its laws to help the believer understand the purpose of existence – they are called dogmas. They are taken on trust, accepted in faith by religious communities and featured in catechisms and theological textbooks. Both science and religion work with conceptual models that propose laws and dogmas which are modified as more about existence is observed and understood. This is demonstrated in the present analysis where the dynamic, indeterminate, relational model of the triune God has similarities with the way quantum physics models particle interactions. Both religious and scientific communities work with these models with considerable success. However, not everyone in the communities is convinced. Some scientists, Einstein among them, cannot accept the indeterminacy of quantum physics and some theologians cannot accept the Trinity because it is "un-Biblical", too much influenced by Greek philosophy. It is important not to place too much weight on these similarities. God is not a quantum event and the indeterminacy of quantum physics does not provide the ultimate gap for the regressive "God of the gaps" argument. What it is possible to say is that the laws of science and the dogmas of religion are important constructs for making sense of the process and purpose of existence, and that the conceptual models that they frame have significant similarities.

The Outworking

If it is accepted that humanity participates in the life of the triune God by trying to understand the process and purpose of existence, formulating laws and dogmas, respectively, what does this involve? The operation of the triune God is to sustain all existence in and through creation and redemption. The requirement of co-inherence is that creating and redeeming must go together. Without redeeming, creating is in danger of falling into the dualist trap. Human beings, both personally and communally committed to the pursuits of science and religion, are co-creators, co-redeemers and co-sustainers, by participating in the life of the triune God. The creative flow that is lovingly given to humanity is to be received and returned through a redemptive spirit. That is when the creative work of humanity is given back in love to deity. What does that mean in practice? It means the giving of one's self to the other, thus reconciling oneself both to another personally and to the many communally. To be selfish as a person or as a community is to receive the given, creative spirit and give nothing back in return, a one-way flow.

Conclusion

In this paper a conceptual model of the triune God has been used to analyse the relationship between the laws of God and the laws of Nature. Drawing on similarities with panentheism, kenotic and process theology, it counters the wide-spread dualist tendency to separate God and the world, the spiritual and the physical, even science and religion. Instead, by bringing the world into the life of the triune God, the laws of science and the dogmas of religion become the way in which scientists and theologians try to understand the process and the purpose of existence.

The conceptual model of God that has been used to analyse the Conference title is obviously a Christian one because it is based on a particular interpretation of the Trinity. However it can be applied whenever the dogma of a belief system trusts in a source beyond the physical and the human. Conversely, it is contrary to systems that eliminate the transcendent (atheism?) or remain in the immanent (humanism?) and/or the connective (popular, "new age" spirituality?).

The theological model used here is a dynamic, indeterminate, relational one and so has some similarities to the models of quantum physics. Without claiming too much for this similarity it does highlight the way in which both scientist and theologian use conceptual models to progress their work.

Finally the way in which the Christian model is worked out in practice has been considered. Both the scientific and religious communities have a common goal of using their creativity to reconcile the self to the other personally, and the one to the many in community. There can be no better task for both communities today.

References

Augustine of Hippo ([399-419] 1991) *The Trinity,* ed.J. Rotelle, transl. E. Hill. Brooklyn, N.Y.: New City Press.

Clayton, P. & Peacocke, A. (eds.) (2004) *In Whom We Live and Move and Have our Being: Panentheistic Reflections on God's Presence in a Scientific World.* Grand Rapids, Michigan: Eerdmans.

Cobb J. B. &Griffin D.R. (1976) *Process Theology: An Introductory Exposition.* Philadephia: Westminster Press.

Emmett, J. C. A. (2001) *Understanding the Trinity in a Scientific Age: An Apologetic Approach.* M.A. Dissertation, Wesley College, Bristol, and Bristol University.

—. (2007) "Kenosis and Creation: Who Empties What into Where and When?" in W.B. Drees, H. Meisinger & T.A. Smedes (eds), *Studies in Science and Theology 11: Humanity, the World and God: Understandings and Actions.* Lund, Sweden: Lund University.

Polkinghorne, J. (2007) *Quantum Physics and Theology: An Unexpected Kinship.* London: SPCK .

Polkinghorne, J. (ed) (2001) *The Work of Love: Creation as Kenosis.* Grand Rapids, MI: Eerdmans, and London: SPCK.

CHAPTER TWELVE

LAW, PRAYER AND THE WORLD'S WEATHER

JOHN LOCKWOOD

Famines

Petitionary prayer is a vast field so I intend to deal with just one aspect, praying about famine and rain. There are over 100 instances of the Hebrew word for "famine" in scripture. Famines are among the most significant natural events reported in the Bible, and have major impacts on many Biblical families. Famines are often, though of course not always, caused by drought: an instance is in the time of Elijah, when 3 years without rain led to famine (1 Kings 17:1-18:45). Famines are often a factor in causing migrations to places where there is more food. Abraham (Genesis 12:10) and Jacob (Genesis 46:5-7) both moved their families to Egypt to avoid famine in Canaan. Ruth 1:1-2 records that Elimelek and Naomi moved to Moab in Transjordan to escape a famine.

Famines, like every other event in nature or history, are integrated into the biblical doctrine of divine providence. It is sometimes claimed that God uses famines as indicators of his displeasure, and as warnings to repent. This was the case in the first of our examples. 1 Kings 16:29-30 reports that King Ahab, more than any of his predecessors, did what was wrong in the eyes of the Lord. Elijah said to Ahab (1 Kings 17:1), "I swear by the life of the Lord the God of Israel..... that there will be neither dew or rain these coming years unless I give the word". Likewise Amos (1:2, 2:6-3:2) ascribes a severe famine, caused by drought and pests, to God's judgement on His rebellious people. This view persists in Revelation 6: 5-8 where famine is a direct visitation on human sin.

Major droughts still occur, but do not necessarily cause major famines. More than half of California is now experiencing the most severe category of dryness and, at the time of writing, experts are warning that the 2015 winter rains are most unlikely to redress the deficit. While this drought has impacted on local agriculture, it is not causing widespread famine in this part of the United States, which has the resources and infrastructure to bring in food from elsewhere. But only a minority of the world's population is able to protect itself from climatic challenges in such a way; if challenges become more widespread, so will the need to migrate to avoid starvation.

Two Stories

Since many famines are caused by lack of rain, prayer about famine is likely to be associated with prayer about rainfall. Rain is caused by weather systems, therefore the question is partly how can prayer influence such a physical system as that of the weather? John Houghton (a Christian and one time Director General of the Met Office) tackles this problem in his paper, "What happens when we pray?" (Houghton 1995a, see also Houghton 1995b). He explains that forecasts of tomorrow's weather are possible because they are dependent on physical processes in the atmosphere that can be described scientifically – in terms of the overall theme of this conference, they are subject to law. So how can he also believe that prayer can have anything to do with it?

Over this time-scale, prayer is indeed unlikely to alter the physical course of events. But it can alter outlooks, both of the person who prays, and of others. It will not so much alter the weather as our response to the weather. Houghton suggests, therefore, that two stories are involved: a "faith story" and a "scientific story". The faith story involves prayer, not just a series of requests, but also the aligning of our will with God's will. It is an attitude of mind where we continually attempt to discover the will of God and act upon it. An example is found in Acts 11:27-29. During this time some prophets came down from Jerusalem to Antioch. One of them, named Agabus, stood up and through the spirit predicted that a severe famine would spread over the entire Roman world. This happened during the reign of Claudius. The disciples, each according to his ability, decided to provide for the brothers living in Judea. This is the first recorded instance of inter-church aid!

The Climate System

To understand the scientific story we need to explore the physics of atmospheric systems. The scientific perspective is difficult because atmospheric processes are complex, non-linear and most probably "chaotic" in the technical sense of being absolutely beyond the possibility of exact prediction (Gleick, 1988; Polkinghorne, 1989, 1994, 1996; Oord,

2010). However the way they behave is not unrestrictedly haphazard, because they explore a confined range of possibilities. Thus chaotic systems exhibit a kind of ordered disorder. The recognition of this behaviour is comparatively recent, since it requires the use of powerful digital computers. In the context of the present conference, we must say that such systems are not "law-governed". These ideas must now be developed.

The climate record shows rapid step-like shifts in climate variability that occur over decades or less, as well as climate extremes (e.g. droughts) that persist for decades. The variability of climate can be expressed in terms of two basic modes: the forced variations which are the response of the climate system (atmosphere, oceans, cryosphere, biosphere) to external influences, and the free variations owing to internal instabilities and feedbacks to non-linear interactions among the various components of the climate system. The external influences operate mostly by causing variations in the amount of solar radiation received or absorbed by the Earth, and comprise variations in both astronomical (e.g. orbital parameters) and terrestrial forcings (e.g. atmospheric composition, aerosol loading). The internal free variations in the climate system are associated with both positive and negative feedback interactions between the atmosphere, oceans, cryosphere and biosphere. These feedbacks lead to instabilities or oscillations of the system on all time-scales, and can operate independently or reinforce external forcings.

What all this means is that the climate system belongs to a particular set of mathematical systems known as *nonlinear systems*. They have unusual, and perhaps unfamiliar, properties that are of great importance to understanding climate change. Indications of non-linearity in the climate system include the following … The Asian monsoon shows an abrupt change or seasonal jump in the northern spring. This abrupt change in the monsoon system is characterized by a sudden shift northward of the subtropical westerly jetstream over Northern India in early June, associated with nearly simultaneous changes in atmospheric circulation parameters over large regions of southern Asia. On a longer time scale, throughout the Alps, the last 3000 years were characterized by repeated glacial events at intervals of 200-400 years, which, in some, were severe enough to have been compared to the Ice Ages (e.g. the "Little Ice Age" of around 1700). The advances were sudden and rapid, with the later ones often obscuring the earlier events. The long-term though sparse records of rainfall in the Sahel – the region south of the Sahara but north of the savannah lands – suggest that conditions of the last couple of decades may be unprecedented in the context of the last several centuries, and also, that shifts from one variability state (e.g. "wet") to another (e.g. "dry") may occur in a couple of years.

The inherent non-linear nature of the atmosphere is manifest in interactions and fluctuations over a wide range of space and time scales. Even on the annual scale, abrupt circulation changes are observed as seasonal changes progress. Many climatic switches take place on time-scales that are relatively short, and make them of significant societal relevance. They are typical of a number of types of behaviour found in dissipative, highly non-linear systems generally, under non-equilibrium conditions, a major instance of which is the climate system. In particular, as a system evolves, due for example to climate change, it may approach a so-called *bifurcation point* where the natural fluctuations become abnormally high, as the system must "choose" among various regimes. These bifurcation points are often known as "tipping points" in the popular literature. Under non-equilibrium conditions, local events have repercussions throughout the whole system, with long-range correlations appearing at the precise point of transition from equilibrium to non-equilibrium conditions. Long-range correlations (teleconnections) are indeed observed in the atmosphere, and the strength of these teleconnections is observed to vary with time.

An important property of developing weather systems or climate systems is that they are markedly sensitive to initial conditions. Slight changes in the initial conditions can lead to very different development pathways. This is a characteristic property of chaotic systems; unless the initial circumstances are known with unlimited accuracy, it is only possible to predict behaviour a small way into the future with any confidence. There is an orderly disorder in their behaviour, because their motions home in on a continual and haphazard exploration of a limited range of possibilities. This feature of chaotic unpredictability first came to light during computer investigations of weather forecasting after the Second World War, and was unexpected. It gave rise to the so-called "Butterfly Effect", the notion that a butterfly stirring the air today in the central Pacific can transform weather patterns next week over London. It makes weather forecasting for more than a few days ahead very difficult. Indeed it places a limit of about 20 days on the range of detailed weather forecasts.

Forecasts are normally produced with the aid of numerical atmospheric models. A standard technique is to vary very slightly the meteorological data used to initiate the model. If the resulting forecasts are very similar, the atmospheric conditions are stable and the weather forecast can be made with some confidence. In contrast if the predictions diverge rapidly, the resulting weather forecasts must be regarded with little confidence. It is this great sensitivity to initial conditions that makes seasonal climate forecasting difficult, if not impossible. Similarly it is not possible to generate detailed climate data for some point in the future; rather the climate prediction has to be expressed in terms of probabilities.

Another interesting property of nonlinear systems is that relatively simple systems can run away into very complex ones. This is possible because of a continual flow of energy into the system from the surrounding environment, for example a warm ocean. Thus scattered cumulus clouds over warm tropical oceans can develop over a matter of days into major tropical storms such as hurricanes.

In the Old Testament we often see people who are reacting, under God's guidance, to climatically caused famines. What is the situation today?

Global Warming

Something started to happen in the years leading up to 1800 AD which is leaving an imprint on both atmospheric composition and temperature not seen for at least the last 800,000, and probably the last 24-26 million years (Houghton 2009; Houghton with Tavner, 2013). An increase of atmospheric Carbon Dioxide concentration started from around 280ppm in 1800, taking it to around 400ppm in 2014. The last time atmospheric carbon dioxide was above 400 ppm was probably about 24-26 million years ago. The cause then is debated, but that starting in the 18th C was the Industrial Revolution, which, for our purposes, effectively began in 1698 with Watt's steam engine and the consequent widespread use of coal. British coal production peaked in 1910, but world production has continued to double every 20 years. From 1769 to 2006, world annual coal production increased 800-fold. Most of this coal was burnt and produced carbon dioxide, a significant amount of which has remained airborne. Much of western prosperity is fossil-fuel based, and is thus responsible for increasing atmospheric carbon dioxide and therefore global warming due to an enhanced greenhouse effect.

Most pollutants released into the atmosphere have very short residence times, often just a few days. Clouds or rainfall wash out the common ones, such as sulphur dioxide, and the larger particles settle out rapidly under gravity. Such pollutants tend to be local in impact. In contrast, the lifetime of anthropogenic carbon dioxide is controlled by the "carbon cycle". According to this, that of the majority could be considered as 300 years, if emissions were to cease tomorrow; but it is considered that, even in that theoretical circumstance of immediate cessation of emissions, 25% would persist for a very long time. The long life of the anthropogenic carbon dioxide released into the atmosphere makes atmospheric stabilization of the gas difficult, and at present nearly impossible in the short term, even for levels below the present atmospheric concentration. About half of cumulative anthropogenic carbon dioxide emissions between 1750 and 2010 have occurred in the last 40 years. Carbon Dioxide concentrations have increased by 40% since pre-industrial times, primarily from fossil fuel emissions. The ocean has absorbed about 30% of the emitted anthropogenic Carbon Dioxide, causing ocean acidification. The pH of ocean surface water has decreased by 0.1 since the beginning of the industrial era: this figure may seem small, but even such a modest change has serious biological implications.

Each of the last three decades has been successively warmer at the Earth's surface than any preceding decade since 1850. This is largely related to the impact of increasing atmospheric carbon dioxide concentration. The average temperature of the Earth's surface has risen by 0.89°C from 1901 to 2012. Ocean warming dominates the increase in energy stored in the climate system. It accounts for more than 90% of the energy accumulated between 1971 and 2010. It is virtually certain that the ocean (0–700 m deep) warmed significantly from 1971 to 2010.

International Action

While the climate meeting in Copenhagen during December 2009 failed to deliver any formal "climate deal", the non-binding Copenhagen Accord recognized the scientific view "that the increase in global temperature above preindustrial levels should be below 2 degrees Celsius". The adoption of this target occurred despite increasing evidence that, for at least some nations and ecosystems, the risk of severe impacts would already be significant at 2°C; hence, the Accord included the intent to consider a lower 1.5°C target in 2015. Therefore the 2009 Accord recognized the scientific view "that the increase in global temperature should be below 2 degrees Celsius", although there were growing fears that this might be too high. At the same time, the continued rise in greenhouse gas emissions in the past decade and the delays in a comprehensive global emissions-reduction agreement have made achieving this target extremely difficult, arguably impossible. Indeed, the horrific possibility of global temperature rises of 3°C or 4°C within this century must now be contemplated.

Climate Change, 2014

In the recently published "Climate Change Synthesis Report" (2014), five reasons for concern have provided a framework for summarizing the key risks associated with climate change. They illustrate the implications of global warming and of adaptation limits for people, economies, and ecosystems across sectors and regions. They provide one starting point for evaluating the dangerous anthropogenic interface with the climate system. They also provide one possible starting point for Christians to consider their responses to future climate-change impacts, particularly as they are partly due to changes in rainfall patterns.

1. Unique and threatened systems: Some ecosystems and cultures are already at risk from climate change. Many systems with limited adaptive capacity, particularly those associated with Arctic sea ice or coral reefs, would be subject to very high risks with warming of 2°C above preindustrial levels.

2. Extreme weather events: Risks from climate-change-related extreme events, such as heat waves and associated, heavy precipitation and associated floods, and coastal flooding, are already appreciable. With 1°C additional warming they would be high.

3. Distribution of impacts: Risks are already unevenly distributed between groups of people and between regions; risks are generally greater for disadvantaged people and communities the world over (Meteorological Office, 2014). Based on projected decreases in regional crop yields and water availability, risks of unevenly distributed impacts will be high if warming exceeds 2°C.

4. Global aggregate impacts: Risks of global aggregate impacts, reflecting impacts on both the Earth's biodiversity and the overall global economy, are moderate under warming of between 1 and 2°C.

5. Large-scale singular events: With increasing warming, some physical and ecological systems are at risk of abrupt and/or irreversible changes. Risks associated with such "tipping points" are moderate between 0 and 1°C warming – there are already signs that both warm-water coral reefs and Arctic ecosystems are experiencing irreversible regime shifts. Risks increase at a steepening rate under an additional warming of 1 to 2°C and become high above 3°C, due to the potential for large and irreversible sea-level rises due to ice sheet loss.

The impacts of climate change are thus a major challenge for the 21st century. One major impact is through changing rainfall patterns causing both floods and droughts. We have seen that in Old Testament times people avoided the impacts of droughts and famines by migrating. In the modern world people are once again choosing this solution to avoid the impacts of droughts (as well as wars!). Increasing numbers of refugees from Africa and the Middle East are trying to enter the European Union. This is increasingly causing major political problems in several European countries, including the UK. As Christians, how do we meet this problem of climate change? Firstly by prayer: we need to look at the faith aspects of the problem and seek God's help, support and direction. Secondly by practical steps, such as using less energy at home and at work, or supporting people living in poor third world countries – this is the scientific aspect of the problem. Christians do therefore need to consider prayerfully their response to global warming.

The Role of Prayer

Writers such as John Polkinghorne (1989, 1996; see also Oord, 2010) have suggested that chaotic systems, such as those determining the weather, could be the loci of God’s interactions with the physical processes of the world, for science could never in principle say that such interactions had occurred. John Houghton (1995a, 1995b) refrains from such explicit suggestions, yet reminds us that God is both outside and within space and time. God is therefore able to act without being constrained by space and time. In any case, the "faith stories" and the "scientific stories" may be different for the same series of events. The possibility of “scientific” proof of the results of prayer is bound to be limited because prayer is a relationship between human persons and God; the tests we apply must be appropriate to a relationship. In any case, for Houghton “the significance in the ‘faith story’ arises from the actions or choices of particular people or from unusualness in the timing or in the sequences of particular events, not so much in the events themselves” (1995a, p. 12). In prayer we cooperate and relate to God in His work in the world, with due regard for the Biblical commands regarding creation care, and seek God’s guidance at every turn in deciding our response.

References

Climate Change Synthesis Report (2014) *IPCC Fifth Assessment Report: Approved Summary for Policymakers, 1 November 2014.*

Gleick, J. (1988) *Chaos: Making a new science*. London: Heinemann.

Houghton, J. (1995a) "What happens when we pray?" *Science and Christian Belief,* 7(1), 3-29.

—. (1995b) *The search for God: can science help?* Oxford: Lion Hudson.

—. (2009) *Global Warming: The Complete Briefing*, 4th edn. Cambridge: Cambridge University Press/

Houghton, J. with Tavner, G. (2013) *In the eye of the storm.* Oxford: Lion Hudson.

Meteorological Office (2014) The "Human dynamics of climate change" poster: http://www.metoffice.gov.uk/climate-guide/climate-change/impacts/human-dynamics

Oord, T.J. (2010) *The Polkinghorne Reader: Science, faith and the search for meaning.* London: SPCK.

Polkinghorne, J. (1989) *Science and Providence.* London: SPCK.

—. (1994) *Science and Christian Belief: Theological reflections of a bottom-up thinker.* London: SPCK.

—. (1996) "Chaos theory and divine action", in *Religion and Science: History, Method and Dialogue,* ed. W.M. Richardson & W.J. Wildman. London: Routledge.

SECTION III

CODA

Chapter Thirteen

"An Undevout Astronomer Is Mad"

Geoffrey Cantor

> The librarian was explaining the benefits of the Dewey decimal system to her junior – benefits that extended to every area of life. It was orderly, like the universe. It had logic. It was dependable. Using it allowed a kind of moral uplift, as one's own chaos was also brought under control.
>
> "Whenever I am troubled," said the librarian, "I think about the Dewey decimal system."
>
> "Then what happens?" asked the junior, rather overawed.
>
> "Then I understand that trouble is just something that has been filed in the wrong place. That is what Jung was explaining of course – as the chaos of our unconscious contents strive to find their rightful place in the index of consciousness."
>
> The junior was silent. I said, "Who is Jung?"

The final "I" in this passage is the young Jeanette Winterson (2011, 127), and this short dialogue appears in her autobiographical novel entitled *Why Be Happy When You Could Be Normal*. In that book she recounted her escape from her crazy, repressive foster mother and discovered new worlds by reading sequentially the books shelved under "English Literature in Prose, A–Z," in the public library at Accrington.

Winterson's librarian lived in a strikingly similar world to that of the eminent theologian and classical scholar Richard Bentley. In one of the early Boyle Lectures, delivered on 5 December 1692, Bentley praised the orderly universe described in Isaac Newton's *Philosophiæ Naturalis Principia Mathematica* (1687). For Bentley, the universe's orderliness demonstrated that it must have been created by God and that the heavenly bodies move according to the laws that God had imposed on matter. Bentley's aim was not only to determine the theological implications of the new Newtonian system but also, as the title of his lecture indicated, to use the perceived order of the universe to confute atheism (Bentley, 1693). He envisaged that the world of the atheist consisted of the uncoordinated movement of Epicurean atoms. Chaos resulted as the atoms zapped around indiscriminately. By contrast, the particles in Newton's divinely created universe moved in an ordered fashion, producing an ordered universe overall. In Bentley's analysis we see the opposition between the God-created universe marked by law and order and the atheist's chaotic world.

This opposition has been employed on many subsequent occasions. For example, in the 1706 edition of his *Opticks*, Newton (1718, 378) likewise contrasted the law-like and God-created universe with the untenable worldview propounded by unphilosophical people who "pretend that it might arise out of a chaos by the mere laws of nature" but without God's designing hand. For Bentley, however, the atheist was not merely unphilosophical but was obviously mad for denying what was patently true. Thus, in an earlier Boyle Lecture he reflected that atheism "is not bare *Folly*, but *Madness* and *Distraction*. … it is manifestly the most pernicious *Folly* and deplorable *Madness* in the *World*" (Bentley, 1692, 35–6). Atheism was a mental illness.

Half a century later the poet Edward Young (1795, 286) reiterated Bentley's characterisation of the atheist when he declared in his widely-read *Night Thoughts*:

> Devotion! Daughter of Astronomy!
> An undevout astronomer is mad.

Madness is manifested by a form of blindness since this irreligious astronomer is unable to see what normal people can see though his telescope. Thus his madness consists in his perverse denial of the divinely designed universe that any sane and rational person could readily see.

The philosophers of the Scottish Enlightenment, who saw themselves as Newton's heirs, used a similar set of opposites. In his *History of Astronomy*, Adam Smith (1795, 27–39) discussed the world of the savage; the savage is a symbolic figure who stands outside civilisation and, like the atheist, is ignorant of Christianity. The savage is at the mercy of nature: He ascribes "thunder and lightning, storms and sunshine, [and] those more irregular events" to the favour or anger of some primitive deity. The savage lives with fear and madness. By contrast, "Philosophy, by representing the invisible chains which bind together all these disjointed objects, endeavours to introduce order into this chaos of jarring and discordant appearances, to allay this tumult of the imagination, and to restore it, when it surveys the great revolutions of the universe, to … tranquillity and composure". Not only is the philosopher/scientist not frightened by thunder and lightning but, understanding its origins, he achieves a mental state of "tranquillity and composure". In contrast to the savage, with his fear-inducing environment, the Enlightenment philosopher, with his superior knowledge, is at ease in the universe.

Some years ago, when I was working on Michael Faraday, I was struck by the importance of law(s) and order both to his science and, more generally, to his life (Cantor, 1991, 201–5). He had a great fear of being overwhelmed and destroyed by chaos, both in the physical world and in society, especially by revolutionary social movements. Thus, in an early commonplace book he noted that an anagram of "Revolution" was "to love ruin". Also, during the period of revolutions that swept across much of Europe in 1848 he wrote to fellow scientists in France and Switzerland expressing the hope that they were safe and could continue their scientific research in peace. In one letter he drew a contrast between the peaceful nature of science and "the crowd of black passions and motives" that impelled the revolutionaries. Like Bentley's atheist, those revolutionaries were perceived as suffering from a mental disorder.

In his own research Faraday sought to confront this apparent chaos in the physical universe, by revealing that it was underpinned by order, as expressed by the laws of nature. As he stated in one of his lectures, the laws of nature were "gifted" by God unto matter and "were established from the beginning"; thus they were "as *old* as creation". While chaos evoked in him a deep feeling of anxiety and insecurity, the discovery of the divinely-ordained laws created a sense of safety. This contrast was admirably captured by the author of a poetic obituary of Faraday that appeared in *Punch* in 1867:

> Till out of seeming Chaos Order grows,
> In ever-widening orbs of Law restrained,
> And the Creation's mighty music flows
> In perfect harmony, serene, sustained. (Anon., 1867)

Faraday thus fits my historical theme that science has been perceived as the discovery of order – the laws of nature – that consigns chaos and madness to flight.

Let me briefly move back to Winterson's account of the librarian at Accrington and her appeal to Melvil Dewey's Decimal System. Without doubt you would rapidly lose your sanity if you had to search miles of library shelving to find a randomly-placed copy of John Hedley Brooke's *Science and Religion* (1991). Just as Bentley argued that Newton brought order to our understanding of the universe and thus preserved sanity, the Accrington librarian considered that Dewey had rescued libraries from chaos and saved generations of librarians and scholars from madness.

One piece of the jigsaw is still missing: In the quoted passage at the beginning of this chapter Jeanette Winterson asked the librarian, 'Who is Jung?' Although much of Jung's opus has been in print for many years, his *Liber Novus*, otherwise known as his *Red Book*, was only published in 1999 (Jung, 1999). In that strange but fascinating work we see Carl Jung in a new light. He started writing this text immediately after his acrimonious parting from Freud in 1913, and it is a record of the personal crisis that ensued and Jung's subsequent resolution. He later claimed that all his mature and distinctive psychological ideas – the theory of archetypes, individuation, etc – had their genesis in this work. Although it is tempting to read the *Red Book* as a prolegomena to his later analytical psychology, it is a narrative detailing the journey of his soul; beginning with the rejection of its current *weltanschauung* and ending with the discovery of a higher and true form of spirituality. For the historian it is a work of Christian mysticism that would not look out of place on a (properly catalogued) library shelf, beside Hildegard of Bingen or Meister Eckhart.

This is not the world of Isaac Newton, Richard Bentley or Adam Smith. It is not the world of science and order; it is alien to the laws of nature and to the kind of natural theology that leads to "tranquillity and composure". Instead Jung left the safety of the shore and was tossed on the sea of chaos. He feared destruction and the annihilation of the self. The word chaos appears on numerous occasions. One entry reads:

> "Everything inside me is in utter disarray. Matters are becoming serious and chaos is approaching."

He even went beyond order and chaos and perceived a higher-order synthesis between the two:

> "You open the gates of the soul to let the dark flood of chaos flow into your order and meaning. If you marry the ordered to the chaos you produce the divine child, the supreme meaning beyond meaning and meaninglessness."

In the midst of this mystical journey there appears a fascinating entry concerning science:

> "My science", wrote Jung, "was the only way I had of extricating myself from that chaos. Otherwise the material would have trapped me in its thicket, strangled me like jungle creepers."

Science, he seems to be saying, has redemptive powers as it saved him from being consumed by the chaos.

The Accrington librarian could not have known Jung's *Red Book,* but nevertheless she recognised that he had something to say about the chaos of our unconscious which we confront on trying to resolve our difficulties; or, in her case, to file correctly the books on the library shelf. Thus this snippet from Jeannette Winterson indirectly addresses some of the issues of this conference volume. Also it chimes with some of my thoughts over the last few years concerning the mental and emotional lives of scientists. What psychological forces draw certain people to a life in science? What does it mean to live a life in science? Is science a deterrent against madness or is it a form of madness itself? For Richard Bentley and many others science brings order to the universe; it quietens our existential fears of being consumed by the destructive chaos; just as the librarian claimed that the Dewey Decimal System quietened her fears.

But isn't there a danger that, by seeing the world as too ordered, too safe and too sanitised our emotional responses to the physical world and to others will be dulled? Carried to an extreme this can lead to emotional deadness, which is itself a form of madness. "Without living emotion there are no viable relationships, and without relationships there is no world", writes the psychoanalyst/historian Robert M. Young (1994, 52). There's more to us – and to life – than science or, indeed, the Dewey Decimal System.

And where does religion fit in? It too has many facets. It can, of course, offer order and safety. Equally it can be the source of an individual's chaos. Both the ordered and the chaotic can contribute significantly to a person's pilgrimage and aid that person's spiritual and mental growth. Religion can also offer the poetry that is much needed in our over-scientised world. One pertinent example is the nature poetry in the Book of Job, where the chaotic moods of nature are brilliantly portrayed. This is insightfully discussed by Tom McLeish (Chap. 2) and in his recent book, *Faith & Wisdom in Science* (2014).

I began with a quotation and will end with one from another writer who has probably never before been quoted at meetings of this Forum. In his book, *God in Search of Man,* the Jewish philosopher and theologian Abraham Joshua Heschel (2009, 45–6) quoted the verses from the Book of Job that have inspired McLeish and so many other people. Just before citing Job, Heschel wrote:

> "Wonder or radical amazement is the chief characteristic of the religious man's attitude towards history and nature. … As civilization advances, the sense of wonder declines. Such decline is an alarming symptom of our state of mind. [Hum]ankind will not perish for want of information; but

> only for want of appreciation. … What we lack is not a will to believe but a will to wonder."

As the contributions to this volume by Eric Priest and Tom McLeish (Chaps 1 & 2) so effectively show, a heightened sense of wonder lies at the heart of the religious appreciation of the universe.

References

Anon. (1867) "Michael Faraday", *Punch*, 53:101.

Bentley, R. (1692) *The Folly of Atheism ... A Sermon Preached at Saint Martin's in the Fields, March the 7th 1691/2, being the First of the Lecture Founded by the Honourable Robert Boyle, Esquire.* London: Printed for H. Mortlock.

—. (1693) *A Confutation of Atheism from the Origin and Frame of the World. A Sermon Preached at St. Mary-le-Bow, December the 5th 1692. Being the Eighth of the Lecture Founded by the Honourable Robert Boyle, Esquire.* London: Printed for H. Mortlock.

Brooke, J. H. (1991) *Science and Religion: Some Historical Perspectives.* Cambridge: Cambridge University Press.

Cantor, G. (1991) *Michael Faraday, Sandemanian and Scientist. A Study of Science and Religion in the 19th Century.* Cambridge: Cambridge University Press.

Heschel, A. J. (2009) *God in Search of Man*. London: Souvenir Press.

Jung, C. G. (1999) *The Red Book. Liber Novus. A Reader's Edition*, ed. S. Shamdasani. New York & London: W. W. Norton.

McLeish, T. (2014) *Faith & Wisdom in Science.* Oxford: Oxford University Press.

Newton, I. (1687) *Philosophiæ Naturalis Principia Mathematica*. London: Joseph Streater.

—. (1718) *Opticks: or, a Treatise of the Reflections, Refractions, Inflections and Colours of Light*. 2nd edition, London: Printed for W. and J. Innys.

Smith, A. (1795) *Essays on Philosophical Subjects*. Dublin: Printed for Messrs Wogan et al.

Winterson, J. (2011) *Why Be Happy When You Could Be Normal*. London: Vintage.

Young, E. (1795) *The Complaint; or, Night Thoughts on Life, Death and Immortality*. New edn, London: Printed for J. Dodsley et al.

Young, R. M. (1994) *Mental Space*. London: Process Press.

CHAPTER FOURTEEN

RESPONSE TO THE CONFERENCE

PAUL BEETHAM

The Conference was rather like the Wisdom considered by some of the speakers in that, whilst it contained excellent separate components, the whole exhibited properties which were more than the sum of the parts. A piecemeal, "reductionist" response is therefore inappropriate. The scope and breadth of the contributions was breath taking, with historians charting the development of reductionist science and its response in Christian apologetics, philosophical challenges to such reductionism and scientists and mathematicians relating careers, in significant scientific and mathematical research, to their own spirituality. Our theology came most strongly from a scientist, in Tom McLeish's paper on science and the wisdom tradition, although John Emmett's Trinitarian short paper gave us a much needed theological challenge too.

It is a pity that Michela Massimi's paper cannot be included in this volume as it gave a stimulating portal to our considerations, although logically it followed from John Henry's paper which traced the concept of laws of nature from Descartes and Newton to modern secular thinkers. With both speakers we can follow the Deist approach of Descartes and the nomic approach of Newton to the concepts of the early Kant. The development from a legalistic concept of the laws of nature to one derived from Wolff's idea of properties grounded in nature, lead naturally to a secular position which requires no divine authority or intervention. A good example of this might be the contrast between Berkeley's tree and Schrödinger's cat. For both Bishop Berkeley and the quantum physicists, existence depends on perception, but for Berkeley God perceives all things whereas for Schrödinger's cat life or death depends on the perception of the observer. Berkeley's position is well captured in the limericks by Ronald Knox and his anonymous respondent (quoted, e.g., by Brown, 1968, pp 64-66):

There once was a man who said, "God
Must think it exceedingly odd
If he finds that this tree
Continues to be
When there's no-one about in the Quad"

REPLY:
Dear Sir,
Your astonishment's odd:
I am always about in the Quad.
And that's why the tree
Will continue to be,
Since observed by
Yours faithfully,
GOD.

Schrödinger's cat on the other hand, without divine observation, remains in uncertain state of life or death until its box is opened and it is witnessed by the presumably human observer. Berkeley, in seeking to prove the existence of God managed to disprove the independent existence of matter. Schrödinger is rumoured to have said, later in life, that he wished that he had never met that cat. In Descartes and Newton we have God and a mechanistic universe, in quantum mechanics we have a probabilistic universe which does not require a divine presence.

Jonathan Topham's paper gave us a wonderful insight into 19th century British apologetics as he described the range of topics covered in the Bridgewater Treatises. They were a hugely popular attempt, by a range of authors pre-eminent in their fields, to relate Christianity to the world presented by the scientific advances of the 19th century. I found it interesting that Darwin began all six editions of *The Origin of Species* with a quote from Whewell's Bridgewater Treatise.

Juuso Loikkanen gave us an interesting contrast with his rather hesitant paper on Intelligent Design, as he attempted to take a middle way in the ID versus evolution controversy. Modern intelligent design theory contrasts starkly with the Bridgewater Treatises, as it places a fundamentalist Christian theology in opposition to neo-Darwinian explanations of biological development. The use that I.D. makes of complex structures is akin to the "God of the gaps" explanations of a previous generation, both of them changing exemplars as scientific explanations are put forward. Mr Loikannen did seem to scout around the Anthropic Principle without mentioning it, but his suggestion that I.D. should be regarded as a philosophical and theological viewpoint seem to imply it. I fear his optimism that I.D. may make a contribution to the science and religion debate is perhaps a little rash, as proponents of intelligent design take a particular theological viewpoint and seek material evidence to support it.

Nancy Cartwright's magnificent Gowland lecture gave those of us rooted in the Newtonian tradition of science a major challenge, and equally undermined a Deist understanding of a legalistic creation. Her Aristotelian proposal leaves little room for an external Deist God and far less for a Theist one, where such a God is external to the creative order. Might one infer pantheism, or perhaps more likely panentheism, as her theological position? Such a position can perhaps lead us to the insights of John Emmett who uses Trinitarian theology to avoid the dualism that Nancy Cartwright undermines. The idea of external laws runs together with the concept of an external cause and regulation of nature. There is a sense in which Nancy Cartwright's rigorous paper left both the scientist and the theologian in me uneasy, as both look for insights deeper than simply that which exists.

John Lockwood's paper on prayer, although commenting that the spiritual should not be seen as separate from the material, surely gives us an emphatic dualism in his distinction between the spiritual consequences of prayer and our inability to determine its physical effects.

Eric Priest's delightful presentation was shot through with the joy of the scientific endeavour. He conveyed the excitement of the scientist seeking answers to profound questions concerning the nature of the universe, and the deep satisfaction in finding answers that are not simply credible but beautiful. For him as for many of us, the scientific endeavour was closely linked to a personal faith. His paper I feel reflects a journey of many scientists with a personal faith for whom the relationship between the two is not an academic debate but a coherent way of life.

Fascinatingly consonant with Prof Priest's paper was Geoffrey Cantor's equally delightful after-dinner speech: the historian of science and the astrophysicist expressed almost identical fundamental outlooks, notwithstanding their very different evidence bases.

A similar sense of excitement was present in Gavin Hitchcock's short paper concerning the laws of mathematics. In his dramatic sketch he conveyed the passions aroused in Victorian academic debate concerning the nature of the laws of mathematics – in particular, those of algebra. The parallels with the debate on scientific laws are striking. Are these laws fundamental aspects of existence, or human constructs? Like many scientists Gavin Hitchcock ends by moving beyond the analytical and finding beauty and goodness in the mathematics – perhaps pointing beyond a dispassionate approach towards our understanding of the nature of our existence.

Fabien Revol's paper on continuous creation gives us a useful prompt in our understanding of creation theology. The concept of God sustaining creation, rather than simply beginning it, has been down played from the time of Newton. Dr Revol's extension of continuous creation from conservation to the source of novelty leads us quite naturally towards the emphasis on Wisdom found in some of our other theological speakers.

Richard Gunton gives us a timely warning concerning the attractive pro-theist argument on "fine tuning". Many who would reject the intelligent design arguments will happily turn to these fine tuning propositions and yet, as Richard Gunton points out, they are essentially the same: each is a "God of the gaps" approach. Fine tuning gives God a particular agency in one part of creation because without God these situations seem improbable. There is a sense in which Dr Gunton's appeal to faith at the end of his paper has an honesty which is not always present in such arguments. He is right that all theories contain presuppositions, and equally correct that those of faith often hide their presuppositions in "scientific" argument.

I feel here that I need to return to the short but very significant paper of John Emmett who gives us perhaps our most incisive theological input. He echoes Karl Barth (1961) in appealing to us to avoid the standard theological dualism prevalent in much modern Christian creation theology, as well as in the secular and pagan belief systems he quotes. It seems to me that he is leading us away from a dualist theology which in many ways parallels the reductionist philosophy in science with which it has been contemporary. Both seem to offer an attractive and simplistic view, and they seem to compete on that basis. John Emmett's Trinitarian theology is inevitably a Christian one but I think its subtleties might well be found in Medieval Jewish mysticism also.

Fraser Watts's short theological and biological paper leads us nicely to what appears to me to be the highpoint of our conference. He charts for us the most recent developments in biology which acknowledge complexity and are moving away from reductionism; and, like John Emmett, he points to a more subtle view of God's action and relationship with creation giving us what he hopes will be a fruitful parallel between interdependence in nature and that found in God.

For me the highpoint of our meeting was the paper by Tom McLeish on the Wisdom tradition in theology and its benefits in relating theology and science. For him, as for many of us, the book of Job has spoken to the scientist in him and it inspired a stimulating and assuring paper. An eminent scientist, Prof McLeish is well versed in interdisciplinary approaches and able to stand back and present the science and faith relationship in a broader context. He gives a much better context for the science and faith dialogue, whose participants often behave as if theirs is the only question and one to be argued in philosophical and sociological isolation. He uses the debates in Job to challenge our assumptions, as did the original authors, and leaves us with questions rather than simplistic answers as does the original book.

I suppose at this point we should consider whether the question that titled the conference has been answered. The title is ambiguous, leaving us to insert the relationship between laws of nature and the laws of God that it questions. Our journey has taken us from the rationalism of Descartes and Newton through 19th century apologetics to the secularisation that they sought to contain. We have had scientists and mathematicians proclaiming the beauty of creation and its order, and its place in personal faith. The concept of natural laws seemed to wither under the searching gaze of modern philosophy, leaving us to question the profundity of a reductionist understanding of creation. We have however been led away from such simplistic approaches to both science and theology and directed to a much more holistic place. Echoing Arthur Peacocke, who ironically was a biochemist, the science which appeared to lead to some of the greatest reductionism, we have been encouraged to explore complexity. It is a complexity where the whole is greater than the sum of its parts and profound properties exist in the complexities. Such an approach bodes well for both science and theology. I would like to share Tom McLeish's optimism that this will give us symbiosis rather than conflict, although I am aware that such conflict often comes not through differences but through parallels and competition for authority.

References

Barth, K. (1961) *Church Dogmatics, Volume 3, The Doctrine of Creation Part 4*. Edinburgh: T&T Clark.

Brown, C. (1968) *Philosophy and the Christian Faith*. London: Intervarsity Press.

INDEX

* = selected entries
n = footnote